PRINCIPLES AND PRACTICE

Springer-Verlag Berlin Heidelberg GmbH

B. R. Jordan (Ed.)

DNA Microarrays: Gene Expression Applications

With 20 Figures

Springer

DR. B.R. JORDAN
Marseille-Génopole
Parc Scientifique de Luminy, Case 901
13288 Marseille, Cedex 9
France

ISBN 978-3-540-41508-4

Library of Congress Cataloging-in-Publication Data
DNA microarrays: gene expression applications / Bertrand Jordan, ed. p. cm
Includes bibliographical references and index.
ISBN 978-3-540-41508-4 ISBN 978-3-642-56517-5 (eBook)
DOI 10.1007/978-3-642-56517-5
1. DNA microarrays–Laboratory manuals. 2. Gene expression–Laboratory manuals
I. Jordan, Bertrand. QP624.5.D726 D634 2001 572.8'65–dc21

http://www.springer.de

Originally published by Springer-Verlag Berlin Heidelberg New York in 2001

Production: PRO EDIT GmbH, 69126 Heidelberg, Germany
Cover design: D&P, 69121 Heidelberg, Germany
Typesetting: AM-productions GmbH, 69168 Wiesloch, Germany
Printed on acid-free paper SPIN 10792625 39/3130Re – 5 4 3 2 1 0

Preface

The aim of this book is to provide in compact form a comprehensive and practical survey of expression measurement using DNA arrays. I have endeavoured to assemble chapters written by scientists who are actual users of the technology and have had to cope with the various practical problems involved in setting up new methods in the laboratory; I believe this is how such a book can be most useful to its readers.

Chapter 1 provides some background on the history of DNA array development; Chapters 2, 3 and 4 focus on various types of microarrays: glass microarrays are described by experienced academic users, while the less-known (but in some situations superior) nylon microarrays are presented by their developers together with some information on nylon macroarrays that are still widely used. Chapter 5 describes in detail the use of oligonucleotide chips in a research laboratory, again with emphasis on practical aspects. The principles and practice of both data acquisition and data mining in the field of expression measurement are presented in Chapter 6 by very experienced authors, together with a wealth of Internet resources that are particularly useful in this fast-moving field. A short last chapter (Chapter 7) attempts to forecast the likely evolution of this field.

All the authors have strived for clarity and insistence on practical aspects; I can only hope that the result will be satisfactory for the reader.

Marseille-Génopole, March 2001 B. R. Jordan

P.S. One vexing point in the terminology of DNA arrays is that the DNA segments bound to the device are called "target" by some and "probe" by others, while the reverse applies to the labelled material prepared from the sample. I have not tried to force a solution to this problem, apart from ensuring consistency within each chapter.

Contents

Chapter 1

DNA Arrays for Expression Measurement: An Historical Perspective

Bertrand R. Jordan

Chapter 2

Expression Profiling with cDNA Microarrays: A User's Perspective and Guide

Sean Grimmond and Andy Greenfield

Chapter 3

cDNA Microarrays on Nylon Membranes with Enzyme Colorimetric Detection
Konan Peck and Yuh-Pyng Sher

Chapter 4

cDNA Macroarrays and Microarrays on Nylon Membranes with Radioactive Detection
Béatrice Loriod, Geneviève Victorero and Catherine Nguyen

Chapter 5

Oligonucleotide Chips for Expression Analysis: Principles and Practical Procedures

PIERRE CASELLAS, ANNICK PELERAUX and SYLVAINE GALIEGUE

Chapter 6

Gene Expression Data Mining and Analysis

ALVIS BRAZMA, ALAN ROBINSON and JAAK VILO

Chapter 7

Future Trends in the Use of DNA Arrays for Expression Measurement

BERTRAND R. JORDAN

CHAPTER 1

DNA Arrays for Expression Measurement: An Historical Perspective

BERTRAND R. JORDAN[1]

[1] Marseille-Génopole, Parc Scientifique de Luminy, Case 901, 13288 Marseille, Cedex 9, France

1
Introduction

Basically, DNA arrays consist of a series of DNA segments regularly arranged on some kind of support, and the expression measurement involves hybridising the whole array with a labelled DNA or RNA sample. The essential feature is parallel processing: in a single experiment, information is obtained on each of the hundreds or thousands of entities present on the array (e.g. whether or not, and at which quantitative level, they hybridise with a given nucleic acid species). It is this parallelism that makes them so important at a time when many megabases of genome sequence and thousands upon thousands of genes need to be analysed.

2
The Forerunners: Colony Filters and Dot Blots

This principle of parallel processing was already implemented in the 1970s. Colony hybridisation (Grunstein and Hogness 1975) was used to search for specific genes among libraries; dot blots (Kafatos et al. 1979) and slot blots allowed homology determination or expression analysis on series of samples, with radioactive labelling in almost all cases (Fig. 1.1). The density of colony filters could be quite high, up to 10,000 on a Petri dish-sized membrane, but these colonies were arranged at random on the membrane since they resulted from plaque lifts or from direct spreading of a transformation reaction onto the membranes. Thus no permanent record of the colonies existed, and the experiments were directed at isolating one or several "positive" clones, the rest being discarded. Dot and slot blots, on the other hand, were done in ordered format, often with 96-trough devices geared to microtitre plates; the DNA (or RNA) solutions were passed through the membrane under conditions conductive to binding, and the resulting microtitre-sized dot blot displaying up to 96 spots was used for expression or homology analysis (Fig. 1.1).

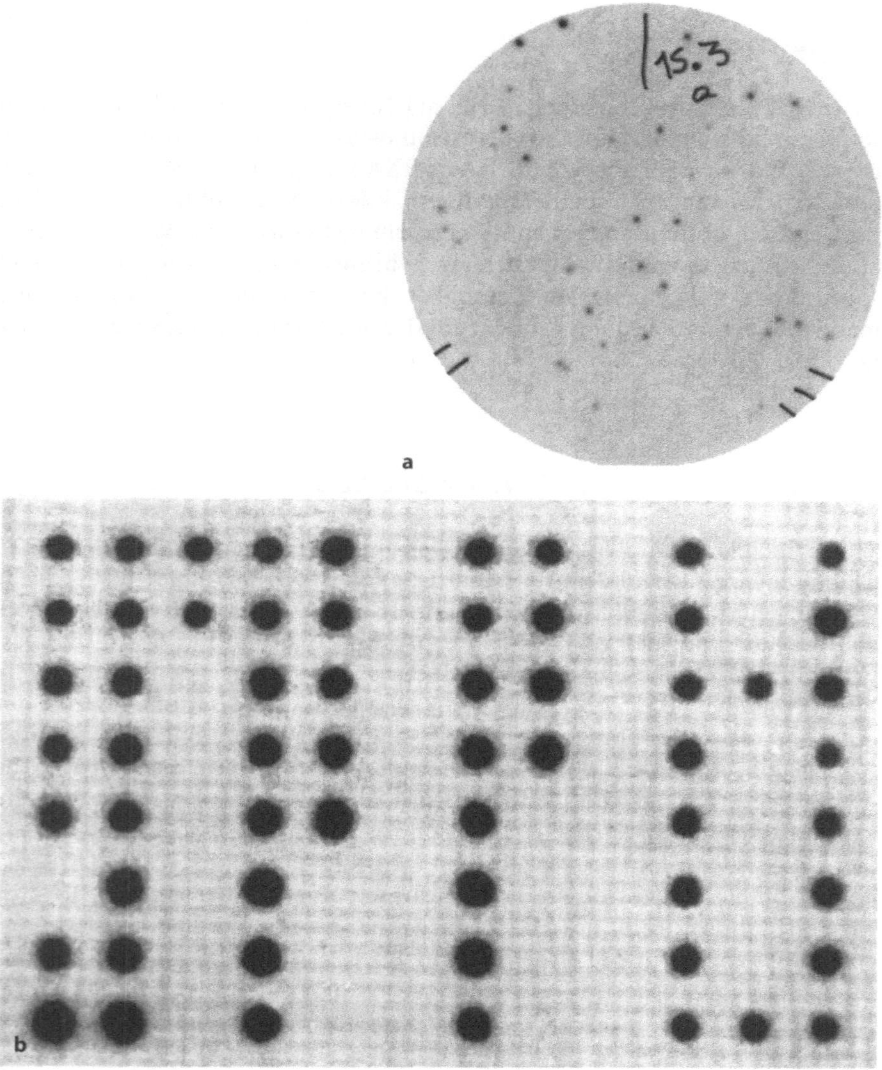

Fig. 1.1. a A phage plaque lift on round nitrocellulose membrane (approx. 1,000 plaques) with a few positive plaques showing up after hybridisation and exposure to X-ray film. Marks on edge of filter are used for subsequent critical positioning of the autoradiograph to recover "positive" plaque material from the Petri dish, since plaques are randomly distributed over the surface. Diameter is approximately 8 cm. **b** An example of the first implementations of ordered DNA arrays, a dot-blot performed with a 96-well filtration device and hybridised with a radioactive probe. Example shown is a screening application with positive/negative samples. Dimensions are ca. 8×12 cm

3
"High-Density" Filters

A major change in the field was the development in the late 1980s of robotic devices ("gridding robots") that made it possible to spot bacterial colonies in a compact and regular pattern. The resulting "high-density filters" contained approximately 10,000 spots on a 22×22 cm^2 surface, corresponding to a "pitch" (centre to centre spacing) of approximately 2 mm (Fig. 1.2). Of course, these colonies had to be ordered previously , i.e. picked from agar plates and distributed into microtitre plates for storage and filter construction. This was often done by hand, although "picking robots" were being developed and reliable commercial models appeared on the market in the early 1990s.

Hans Lehrach's group, at EMBL (Heidelberg, Germany), then at ICRF (London, UK) and now at the Berlin Max Planck Institute, was the major proponent of this approach for genome analysis (Hoheisel et al. 1991; Lennon and Lehrach 1991), at a time when the approach taken in the USA relied almost exclusively on PCR methods and PCR screening of cleverly arranged pools of clones (Olson et al. 1989; Green and Olson 1990) – hybridisation being considered "not robust enough" for systematic genome work. Lehrach and coworkers introduced the concept of the "reference library" as a defined and more or less permanent set of clones, each of which has a definite physical address (e.g. well B12 of plate 312), is present at a known position in a set of high-density membranes, and is stored with all the information obtained in a relational database (Nizetic et al. 1991). This made it possible to acquire, store and correlate data from many different laboratories, as long as they used the same membranes and sent back their results to the central database, a major hurdle in practice.

For a while, high-density membranes were mostly used for library access. Several resource centres accumulated various genomic and cDNA libraries and distributed them to laboratories in the form of high-density membranes. When the user had identified a positive spot using a probe for a gene of interest, the centre then provided the corresponding clone (Fig. 1.2). This approach worked fairly well, at a time when sequence information was very scarce, and allowed many users to progress much faster with their projects. As noted before, information feedback to the resource centre has always proved difficult to enforce; the data coming from very diverse laboratories also turned out to be difficult to handle, so that, contrary to early expectations, this approach did not contribute much to the construction of large-scale physical maps.

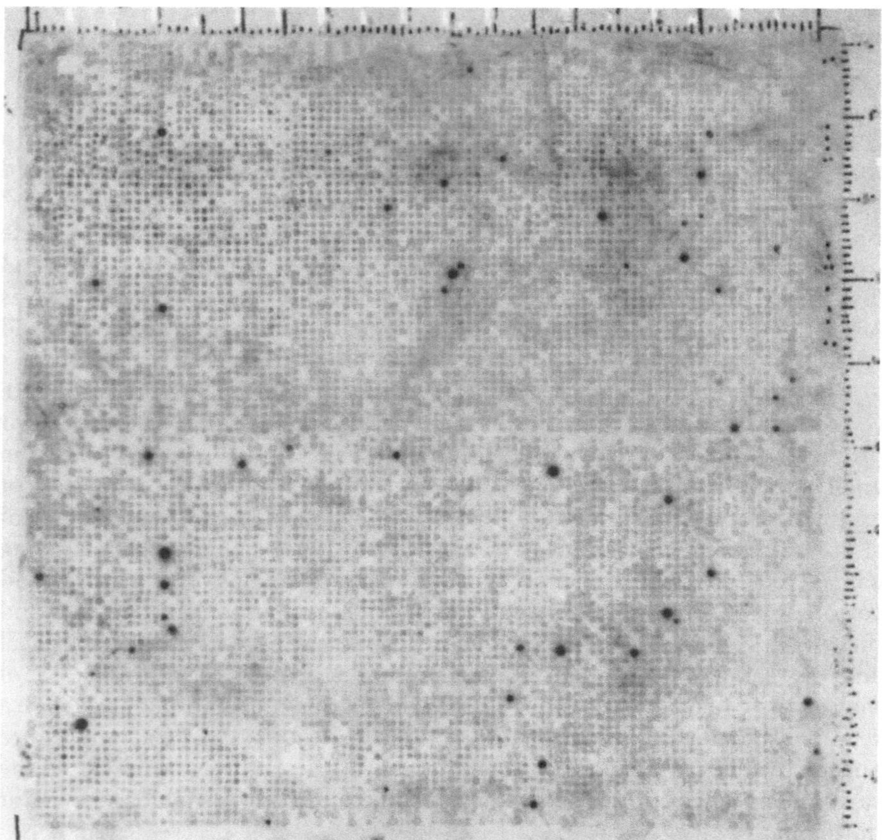

Fig. 1.2. An early "high-density filter" used to access genomic or cDNA libraries. In this case, 10,000 bacterial colonies containing DNA segments cloned in cosmids have been spotted and grown on a 22×22 cm nylon membrane. Hybridisation with a probe prepared from the insert of a cDNA clone reveals one or several positive spots, indicating the corresponding genomic clone; the relevant cosmid can be ordered from the resource centre that provided the membrane by indicating its coordinates. The whole grid is visible because of light background hybridisation; successive hybridisations to search for various genomic clones are performed on the same membrane without stripping (to avoid loss of material), hence the large number of "positive" spots. This was the first widely used application of high-density DNA arrays, pioneered by Hans Lehrach's group and involving only positive/negative scoring, usually by visual inspection

4
Colony Filters and Expression Analysis (Qualitative)

The use of unordered or ordered colony filters containing cDNA clones for expression analysis began in the early 1980s with a qualitative application, differential screening. Parallel hybridisation of "identical" membranes containing clones from conventional or subtracted cDNA libraries with complex labelled cDNA mixtures prepared from two different samples was used, alone or in conjunction with other methods, to pinpoint genes whose expression was different under the two conditions. Using radioactivity and X-ray film detection, this method was necessarily qualitative; it was also cumbersome and suffered from a number of technical problems, but was nevertheless instrumental in isolating several important genes such as the T cell receptor (Hedrick et al. 1984) or the CTLA (cytotoxic T lymphocyte-associated transcripts) series of molecules (Brunet at al. 1988).

5
Quantifying Expression with High-Density Filters

The concept of using the newly developed imaging plate systems that were just beginning to penetrate biological research laboratories (Amemiya and Miyahara 1988) for quantitative acquisition of these data and more refined analysis of expression patterns was fairly obvious and discussed as early as 1990, with a first publication reporting actual data in 1992 (Gress et al. 1992), but the full implementation of the technology took a fair amount of time. This is not unusual when moving from a qualitative to a quantitative application, especially as expression measurement with DNA arrays involves the quantification of very weak signals if data for genes expressed at very low levels are required. A number of artefacts leading to spurious data (especially with the unsequenced cDNA libraries of that period) had to be identified and taken care of (Nguyen et al. 1995), standardisation methods had to be worked out (Bernard et al. 1996), and the rather primitive and user-unfriendly image analysis programs of the period had to be improved (Granjeaud et al. 1996) – a task made easier by the quickly increasing computing power available to scientists. Automation of PCR and standardisation of plasmid vectors used for cDNA libraries led to a shift from colony filters to membranes on which amplified DNA has been deposited, although colony filters are still used in some cases.

During the first half of the 1990s, a few groups worked out the methods, published proof-of-principle papers (Nguyen et al. 1995; Zhao et al. 1995; Pietu et al. 1996) and began accumulating expression data in different systems (Fig. 1.3), leading later to biological results published in the relevant journals. Some of this work was performed in "gene discovery" mode, i.e. measuring the expression level in a number of conditions for a large set of genes (represented by sequenced or, in some cases, unsequenced cDNAs), in order to pinpoint genes whose expression patterns suggest their involvement in particular biological events (see, e.g., Carrier et al. 1999). In this type of approach, extensive biological knowledge is necessary at the start, to define precisely the conditions that are likely to yield the most significant information, and later to choose the "best" genes out of the dozens that display the required expression profile and to define and perform the additional experiments that will allow this choice to be done most effectively. In this context, the biological samples are used to obtain information on the genes and to highlight the most interesting ones (preferably "new", i.e. not yet described in detail), whose close study is likely to be the most rewarding with regard to the biological question approached.

Expression measurement was also performed in an "expression profiling" mode, in which the set of genes (often more restricted) was chosen a priori and usually well known, and the objective was to obtain information on the samples: analysis of the expression profile for a series of tumours, for instance, in the hope of obtaining prognostic and therapeutic information (see, e.g., Bertucci et al. 1999b). In this case the genes are used as tools to derive information on the samples, exactly the opposite of the previous situation. Of course, if technical progress, miniaturisation and decreasing costs make it possible to routinely include all human genes in every experiment, the two approaches will eventually coalesce.

The high-density filter (under the more trendy name of "macroarray") is still with us: for experiments of moderate scope, it performs quite adequately and blends well with current laboratory practice and equipment. In fact, a number of manufacturers market such membranes with sets of cDNA clones (in the form of PCR products or bacterial colonies) from various organisms, an indication of the continuing popularity of the method. Even a firm like Incyte, which has heavily promoted the microarray approach after acquiring a specialist company, Synteni, now also sells macroarrays.

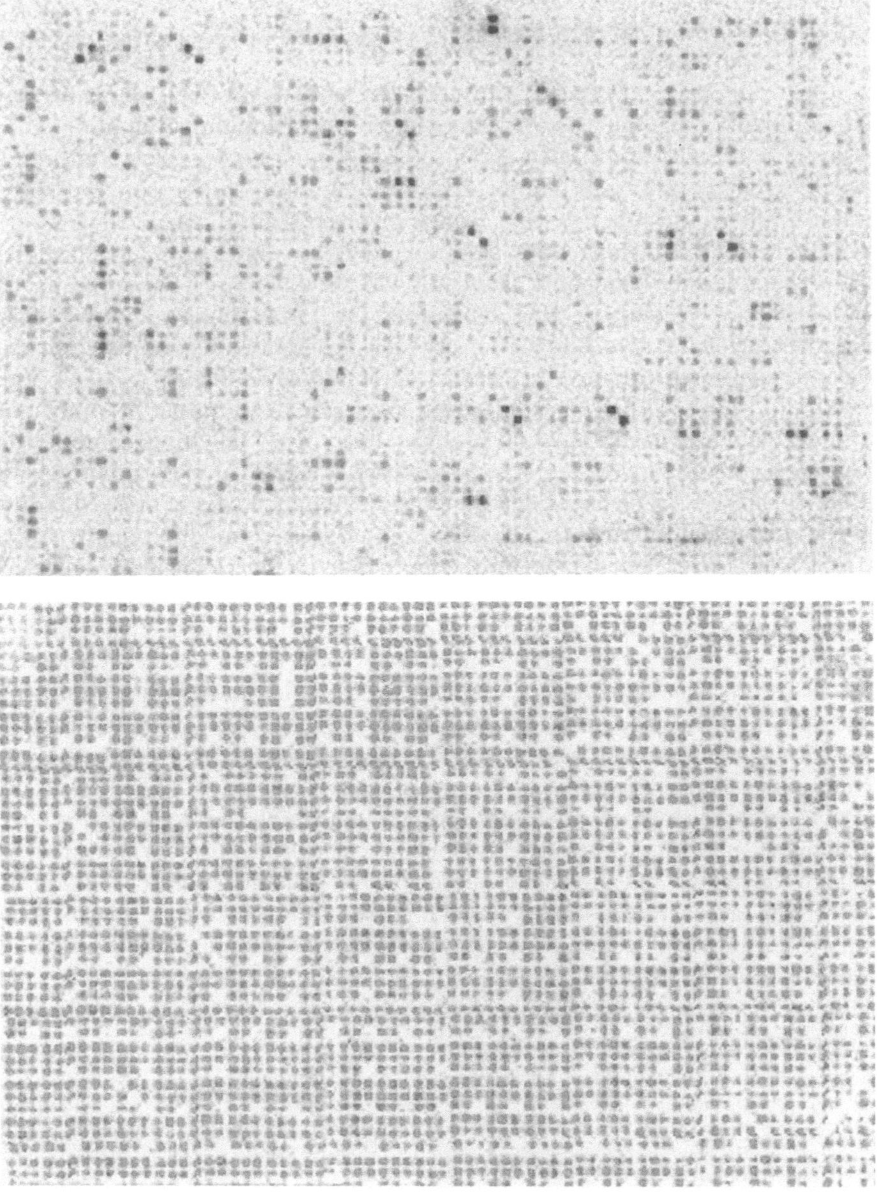

Fig. 1.3. Typical macroarray on nylon membrane, in this case measuring 8×12 cm and containing 3,072 bacterial colonies. *Bottom* hybridisation with a common vector sequence showing all spots (and allowing measurement of relative amount of DNA in each of them); *top* hybridisation with radioactive cDNA made from total mRNA. Hybridisation intensities are acquired and quantified using an imaging plate system, allowing measurement of relative mRNA abundances

6 Miniaturisation: cDNA Microarrays

In the mid-1990s, miniaturisation became a major issue in the further development of DNA arrays, with the aim of increasing the number of genes assayed in a single experiment, and also of reducing sample usage - although most current systems still require microgram amounts of messenger RNA, a major limitation in practice. One avenue involved scaling down cDNA arrays to a pitch (centre to centre spot distance) of 200 to 400 μ (up to 2,500 spots/cm^2). This was done using optical detection methods (fluorescence) because of their superior resolution, and depositing the DNA spots on very planar supports (glass slides) to allow intensity measurement with confocal optics in order to achieve the required sensitivity. First published in 1995 (Schena et al. 1995), this approach has blossomed and given rise to a number of important studies. The use of fluorescent probes allows dual labelling, simplifying comparisons and facilitating standardisation of series of experiments; very good sensitivity (in terms of detection of messenger RNAs expressed at low levels) has been obtained, although sample requirements remain high. Microarrays can be constructed in the laboratory; the necessary equipment is now commercially available, although the expense and logistics are not trivial. Ready-made arrays have now appeared on the market; this has been a relatively slow process due not only to the time taken to build up the necessary logistics but also to intellectual property issues. Generic microarrays should also be made available by academic resource centres. Chapter 2 gives an overview of this important approach and a concrete description of its implementation from a user's perspective.

Nylon microarrays can also be produced. Because of the intrinsic fluorescence of all nylon supports (so far), detection must be performed by enzymatic means that are convenient and affordable, but relatively insensitive (Chen et al. 1998), or with (33P) radioactive labelling, using high-resolution detectors that provide sufficient resolution to quantify arrays with 400-μ pitch. In this form the method makes expression profiling at reasonable sensitivity with very small biological samples (Bertucci et al. 1999a) possible. These two facets are described in Chapters 3 and 4.

7 Miniaturisation: Oligonucleotide Chips

The other, competing approach is that of oligonucleotide chips, pioneered by S. Fodor and the firm Affymetrix. These glass chips, carrying hundreds of thousands of small (24×24 μ) "features", each containing several million copies of a given oligonucleotide (20- to 25-mer), were originally developed for "quasi-sequencing" (mutation detection) applications (Fodor et al. 1991).

They have been shown to allow expression measurement as well, at the expense of assaying each gene with several (20–40) oligonucleotides and controls, in order to average out signal and background artefacts due to the vagaries of short oligonucleotide hybridisation (Wodicka et al. 1997). Their manufacturing process, very similar to that of electronic devices, promises further miniaturisation beyond the present 400,000 feature chip. Being based solely on sequence knowledge, they do not require the cumbersome logistics of cDNA clone storage and PCR amplification, in contrast to cDNA arrays; however, the approach lacks flexibility, the chips are very expensive and their use in the academic sector is still rather limited. Chapter 5 provides a description of the practice of using these chips in an (industrial) research laboratory. Alternative approaches to oligonucleotide chips (notably different synthesis methods) may change the outlook and make this approach more flexible and user-friendly.

8
Data Acquisition, Storage and Mining

The importance of software issues in expression measurement was not immediately recognised, but quickly became apparent as more and more massive data began to flow from these experiments. It includes a number of aspects, from verification of the validity of measurements to sophisticated data mining through data representation and storage issues. Although available computing power has increased dramatically in the last decade, these issues are still far from being satisfactorily solved. They are tackled in Chapter 6.

9
A (Provisional) Conclusion

It is certain that expression measurement by DNA array methods will continue to be an important component of most genomics projects, for a very simple reason: it is, and will remain for some time, the only method that can add some functional information to thousands of sequences, with a throughput of the same order as that of DNA sequencing. However, it is not easy to forecast the directions in which this field will develop. The technology is still extremely fluid and new developments may have strong and short-term impact. In addition, as this field has been heavily invested by industry, commercial imperatives lead to scanty and overoptimistic communication which sometimes makes it difficult to foresee rationally future trends. I have, however, tried to sketch likely developments in the last chapter of this book (Chapter 7).

References

Amemiya Y, Miyahara J (1988) Imaging plate illuminates many fields. Nature 336:89–90

Bernard, K, Auphan, N, Granjeaud, S, Victorero, G, Schmitt-Verhulst AM, Jordan BR, Nguyen C (1996) Multiplex messenger assay: simultaneous, quantitative measurement of expression for many genes in the context of T cell activation. Nucleic Acids Res 24:1435–1443

Bertucci F, Bernard K, Loriod B, Chang YC, Granjeaud S, Birnbaum D, Nguyen C, Peck K, Jordan BR (1999a) Sensitivity issues in DNA array-based expression measurements and performance of nylon microarrays for small samples. Hum Mol Genet 8:1715–1722

Bertucci F, Van Hulst S, Bernard K, Loriod B, Granjeaud S, Tagett S, Starkey M, Nguyen C, Jordan B, Birnbaum D (1999b) Expression scanning of an array of growth control genes in human tumor cell lines. Oncogene 18:3905–3912

Brunet JF, Denizot F, Golstein P (1988) A differential molecular biology search for genes preferentially expressed in functional T lymphocytes: the CTLA genes. Immunol Rev 103:21–36

Carrier A, Nguyen C, Victorero G, Granjeaud S, Rocha D, Bernard K, Miazek A, Ferrier P, Malissen M, Naquet P, Malissen B, Jordan BR (1999) Differential gene expression in CD3e- and RAG-1 deficient thymuses: definition of a set of genes potentially involved in thymocyte maturation. Immunogenetics 50:255–270

Chen JJ, Wu R, Yang PC, Huang JY, Sher YP, Han MH, Kao WC, Lee PJ, Chiu TF, Chang F, Chu YW, Wu CW, Peck K (1998) Profiling expression patterns and isolating differentially expressed genes by cDNA microarray system with colorimetry detection. Genomics 51:313–324

Fodor SP, Read JL, Pirrung MC, Stryer L, Lu AT, Solas D (1991) Light-directed, spatially addressable parallel chemical synthesis. Science 251:767–773

Granjeaud S, Nguyen C, Rocha D, Luton R, Jordan BR (1996) From hybridization image to numerical values: a practical, high throughput quantification system for high density filter hybridizations. Genet Anal Biomol Eng 12:151–162

Green ED, Olson MV (1990) Systematic screening of yeast artificial-chromosome libraries by use of the polymerase chain reaction. Proc Natl Acad Sci USA 87:1213–1217

Gress TM, Hoheisel JD, Lennon GG, Zehetner G, Lehrach H (1992) Hybridization fingerprinting of high-density cDNA-library arrays with cDNA pools derived from whole tissues. Mamm Genome 3:609–619

Grunstein M, Hogness DS (1975) Colony hybridization: a method for the isolation of cloned DNAs that contain a specific gene. Proc Natl Acad Sci USA 72:3961–3965

Hedrick SM, Cohen DI, Nielsen EA, Davis MM (1984) Isolation of cDNA clones encoding T cell-specific membrane-associated proteins. Nature 14(308):149–153

Hoheisel JD, Lennon GG, Zehetner G, Lehrach H (1991) Use of high coverage reference libraries of *Drosophila melanogaster* for relational data analysis. A step towards mapping and sequencing of the genome. J Mol Biol 220:903–914

Kafatos FC, Jones CW, Efstratiadis A (1979) Determination of nucleic acid sequence homologies and relative concentrations by a dot hybridization procedure. Nucleic Acids Res 7:1541–1552

Lennon GG, Lehrach H (1991) Hybridization analyses of arrayed cDNA libraries. Trends Genet 7:314–317

Nguyen C, Rocha D, Granjeaud S, Baldit M, Bernard K, Naquet P, Jordan BR (1995) Differential gene expression in the murine thymus assayed by quantitative hybridization of arrayed cDNA clones. Genomics 29:207–215

Nizetic D, Zehetner G, Monaco AP, Gellen L, Young BD, Lehrach H (1991) Construction, arraying, and high-density screening of large insert libraries of human chromosomes X and 21: their potential use as reference libraries. Proc Natl Acad Sci USA 88:3233–3237

Olson M, Hood L, Cantor C, Botstein D (1989) A common language for physical mapping of the human genome. Science 245:1434–1435

Pietu G, Alibert O, Guichard V, Lamy B, Bois F, Leroy E, Mariage-Sampson R, Houlgatte R, Soularue P, Auffray C (1996) Novel gene transcripts preferentially expressed in human muscles revealed by quantitative hybridization of a high density cDNA array. Genome Res 6:492–503

Schena M, Shalon D, Davis RW, Brown PO (1995) Quantitative monitoring of gene expression patterns with a complementary DNA microarray. Science 270:467–470

Wodicka L, Dong H, Mittmann M, Ho MH, Lockhart DJ (1997) Genome-wide expression monitoring in *Saccharomyces cerevisiae*. Nat Biotechnol 15:1359–1367

Zhao N, Hashida H, Takahashi N, Misumi Y, Sakaki Y (1995) High-density cDNA filter analysis: a novel approach for large-scale, quantitative analysis of gene expression. Gene 156:207–213

CHAPTER 2

Expression Profiling with cDNA Microarrays: A User's Perspective and Guide

Sean Grimmond[1] and Andy Greenfield[2]

[1] Institute of Molecular Bioscience, University of Queensland, St. Lucia, Queensland 4072, Australia
[2] MRC Mammalian Genetics Unit, Harwell, Didcot, Oxfordshire OX11 0RD, UK

1 Introduction

Genomic science has now advanced to the point where it is possible to define the genomic structure and gene content of any organism. Such advances have led to the development of expression tools that can study gene expression in a massively parallel fashion. These methods are also affordable, sensitive, discriminating, and require minimal sample RNA. cDNA microarray expression profiling is a rapidly developing technology that makes possible monitoring of gene expression on a genome-wide scale.

When RNA source is not limiting expression profiling can reach sensitivities of detection as high as 1/500,000 transcripts (Schena et al. 1996). The discriminating power of microarrays is similar to that of other hybridisation techniques (such as Southern or Northern hybridisations). When performed under stringent conditions it is possible to discriminate between elements that possess less than 85% nucleotide homology (DeRisi et al. 1996). Techniques have been established to make RNA target from minute amounts of RNA (Dixon et al. 1998; Lou et al. 2000) and arrays containing up to 18,000 elements have been reported (Alizadeh et al. 2000). The true power of this technology was demonstrated by expression analysis of the entire gene complement of *Saccharomyces cerevisiae* (DeRisi et al. 1997). The challenge now is to exploit DNA microarrays for expression profiling of more complex systems such as mammalian cell lines and tissue samples.

2 Expression Profiling with cDNA Microarrays: The Basics

The use of cDNA microarrays involves three stages that are summarised in Fig. 2.1. The first stage (Fig, 2.1A) involves the preparation, arraying and attachment of DNA probes (also known as elements) to a non-porous substrate. The DNA elements used to make expression microarrays are normally PCR products amplified from cDNA, using either gene- or vector-specific oligonucleotides. The non-porous substrate onto which the DNA probes are arrayed is typically a treated glass slide whose surface has been modified to bind DNA (traditionally coated with poly-L-lysine (Schena et al. 1995). A robotic arrayer is used to print out the DNA elements at very high density and the arrayed DNA is fixed to the surface.

The second stage of expression profiling involves preparation of labelled cDNA pools (known as labelled target) from a test and reference RNA sample. Each sample is labelled using a different fluorescently labelled nucleotide (e.g. Cy5-dCTP for reference, Cy3-dCTP for test RNA). Both labelled populations are then pooled and co-hybridised to the same cDNA array.

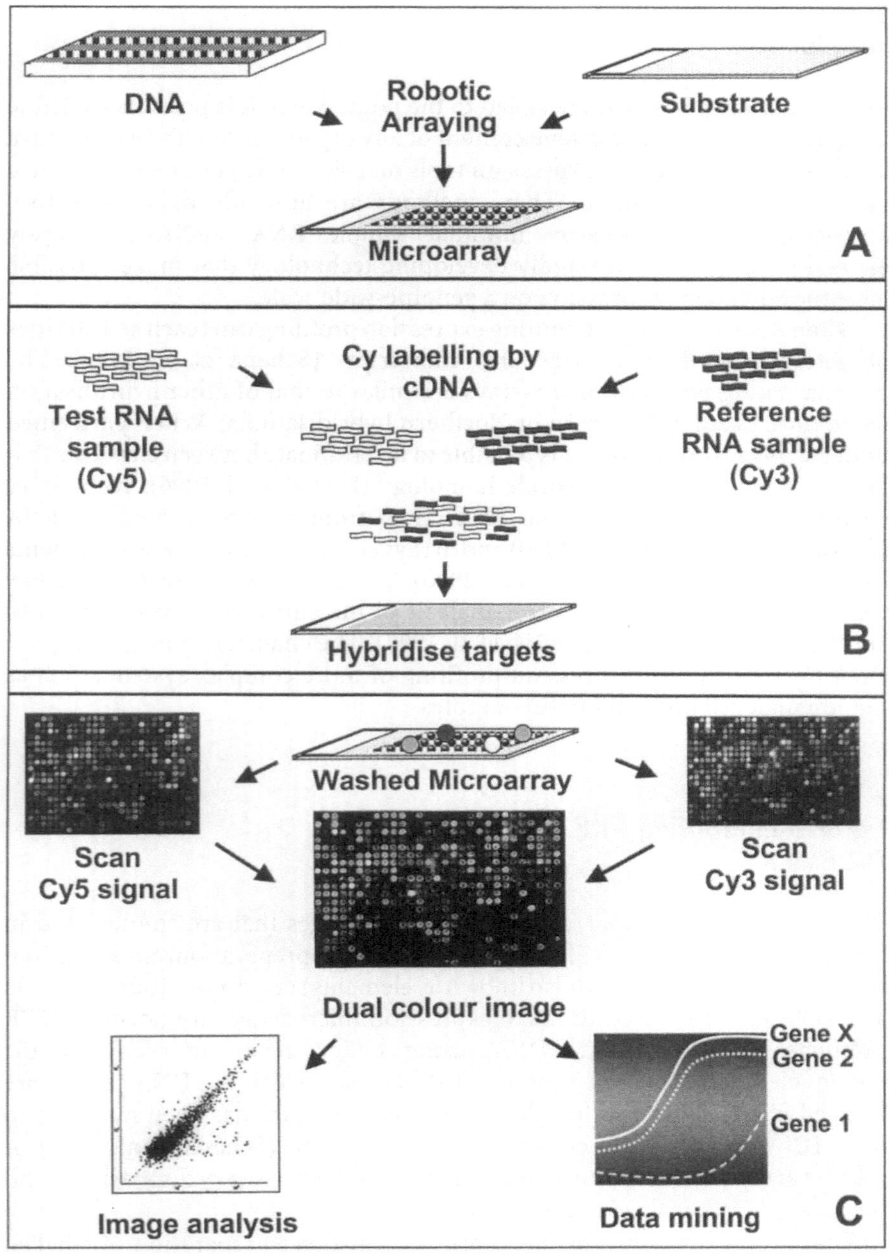
DNA
Robotic Arraying
Substrate
Microarray
A
Cy labelling by cDNA
Test RNA sample (Cy5)
Reference RNA sample (Cy3)
Hybridise targets
B
Washed Microarray
Scan Cy5 signal
Scan Cy3 signal
Dual colour image
Gene X
Gene 2
Gene 1
Image analysis
Data mining
C

After hybridisation and washing, the third stage involves quantifying the test and reference signals of each fluorophore for each element on the array, traditionally achieved by confocal laser scanning. Image analysis software is used to determine signal for each and the differentially expressed genes are identified.

Even at this early stage in the development of microarray technology, many laboratories worldwide are looking to establish expression profiling facilities. The aim of the rest of this chapter is to describe the practical aspects of microarray experimental design and certain considerations worth making prior to establishing a DNA microarray expression system. The material in this chapter is based predominantly on our own experiences of constructing and using cDNA microarrays with commercially available hardware and software. A conventional description of our methodologies and data can be found in Grimmond et al. (2000).

3 Production of cDNA Microarrays

3.1 Experimental Design

Expression profiling using cDNA microarrays has the potential to identify small differences in gene expression levels (in the order of twofold) between any two biological samples. This level of sensitivity and resolution can create problems of interpretation if experiments are not designed carefully. The following are general considerations for any expression profiling experiment.

3.1.1 Minimise Transcriptional Consequences Unrelated to the Biology

It is essential that the test and reference samples be carefully chosen to minimise the detection of differences that are unrelated to the biology of interest. It is important to exclude as much inter-sample variation as possible, from the culturing and sampling of the tissues to the preparation of RNA. This requires careful monitoring of reagents and plastic-ware to ensure that the same batches

Fig. 2.1. Three part schematic showing stages of an expression profiling experiment. A Microarray manufacture involves spotting of DNA elements onto specially coated glass slides using robotic delivery system. B Target labelling involves preparing fluorescently labelled cDNA pools from test and reference RNA samples and then hybridising both targets to same microarray. C After hybridisation, array is washed and then scanned with a confocal laser scan or CCD camera to determine amount of test and reference fluorescence hybridised to every element on the array. These data are often displayed as a dual colour image to allow for rapid identification of differentially expressed genes. Computer software is then used to correct true signal values and provide relative quantification levels of gene expression and can ultimately be mined using bio-informatic tools

are used throughout when inter-experiment comparison is required. In the case of the analysis of mouse models, genetic background differences between wild type and mutant mouse strains should be considered. It is unclear how many significant environmental factors may exist to confound a comparative analysis – the ambient temperature, humidity, time of day and individual researcher involved at the time of sampling may all affect the expression profiles generated.

3.1.2 Collect Appropriate Samples for Analyses

Firstly, it is always important to focus on target tissues that are central to the biological questions being addressed. Poor choice of tissue sample can result in little data of value, because critical transcriptional differences either are not present in the samples chosen or are diluted out by contaminating cell types. In the case of expression profiling of tumours, stromal cell contamination can vary widely between individual tumour samples. If total RNA is extracted from whole solid tumour samples and expression profiles compared, this variation in degree of stromal contamination will be reflected as expression differences. To avoid these sorts of problems, some researchers have resorted to collecting individual critical cell types from tissue samples via cell-sorting techniques such as laser capture microscopy (LCM) (Kacharmina et al. 1999).

When attempting to identify the transcriptional differences between reference and test samples, it is crucial that the test sample be collected at a suitable time-point, e.g. at a certain time after a particular challenge. (Challenges may range from a mutation to the administration of a particular pharmaceutical agent.) If samples are collected too soon, important biological responses may be missed and few transcriptional differences will be observed. If the time-point selected is too late, it is likely that many of the differences observed will be unrelated to the specific challenge but will simply reflect general consequences way downstream. To circumvent these problems, it is always preferable to perform a series of experiments at different time-points to assess expression profiles. Such a time-course of expression profiles also facilitates clustering of genes into particular functional groups that may shed light on the specific pathways affected by a challenge.

3.1.3 Use Elements Relevant to the Biology

Even if one manages to collect the most biologically relevant material to examine, the informativeness of the data generated is only as good as the elements that are interrogated. If the identification of genes activated in response to a given stimulus is the objective, elements generated from normal resting tissues are unlikely to contain relevant probes. To this end many researchers have attempted to generate cDNA libraries which are most relevant to the systems they wish to study. Conventional cDNA libraries are not normally used without some adjustment of the level of abundant classes of elements by normalisation or subtraction (Soares et al. 1994; Bonaldo et al. 1996).

The recent availability of large minimal sets of cDNA clones (e.g. the RIKEN and DFKZ minimal cDNA sets and the commercially available sets (http://www.riken.go.jp/eng/index.html, http://www.dkfz-heidelberg.de/abt0840) has meant that a genomics-oriented approach is now possible. Although such sets are not biased towards a particular biological system, they have the advantage of providing minimal element redundancy. The logical conclusion of this approach is the complete genome set, and it is likely that these will be available in the not too distant future. The challenge will then be to increase probe density, allowing a complete genome set to be printed as a single array.

3.1.4 Always Reduce Complexity

Expression studies on cell lines are capable of detecting less than one transcript per cell when labelled target is not limiting. As the transcriptional complexity of an RNA sample increases, however, the overall sensitivity of an array experiment decreases. This becomes an issue when attempting to perform expression profiling on whole organs or tissue samples, but can be circumvented with the use of specific cell-capture methodologies allied to RNA amplification protocols (Kacharmina et al. 1999).

3.2 Microarray Construction

3.2.1 Practical Tips

A microarray experiment is only as good as the arrays used to perform the expression profiling. Extreme care should be taken when preparing both substrates and elements and the following tips highlight common sources of background on arrays. Wearing non-latex gloves and handling slides with forceps minimises exposure of substrates to oil and talc (common sources of non-specific background hybridisation). Solutions should be filtered to remove dust or particulate material. Frosted slides should be avoided as the white paint is often distributed across the surface of the substrate. Slides should always be spun dry at ambient temperature (500 rpm for 5 min) to prevent watermarks on the arrays. Watermarks can give local background that may not appear until hybridisation.

3.2.2 Elements

Elements can be divided into two classes: experimental and controls. In the case of cDNA microarrays, the experimental class of elements consists of PCR products generated from cDNAs whose expression profile is unknown. The number of these clones varies from study to study, but they usually comprise the bulk of elements on any given array.

As in all scientific experimentation, the control class of elements is essential to gauge the success and reliability of any given data set. In the case of establishing expression profiling, controls are vital to rapidly identifying problem areas in the methodology. Control elements can be broken down into three groups: (1) negative controls, (2) positive controls and (3) key controls.

1. The negative control class comprises a set of probes designed to detect non-specific and artefactual hybridisation signals. Typically we include a series of non-mammalian probes (plant or bacterial cDNAs) that should not detect any specific signal. A poly A tract probe is included to detect poor blocking of oligo dT tails found in labelled cDNA target, and a CoT1 DNA probe is included to detect non-specific signal associated with poor blocking of repeat sequences. Care should be taken to precipitate CoT1 DNA prior to arraying because commercial sources of CoT1 DNA supply the material in a Tris-buffered solution that prevents efficient attachment of the DNA to the treated glass substrate. Blank elements of arraying solution should also be included to confirm that probe carry-over is not taking place.
2. The positive controls used in our system were designed to evaluate the success of different stages of the expression profiling process. Firstly, fluorescently labelled cDNA probe was arrayed onto key positions of the array to act both as a positional guide and as a control to measure element retention throughout the entire procedure, from array spotting to post-hybridisation washes. Figure 2.2 shows the tracking of element retention on poly-L-lysine-coated slides by the use of printed fluorescently labelled probes. One disadvantage of this sort of control element is that it is prone to bleaching over time. It is important to note that the arraying of fluorescently labelled PCR products is not recommended for arraying systems that use quill-type pins because probe material can coat the pin and result in carry-over. All our arraying was performed using a GMS/Affymetrix 417 microarrayer that uses a solid pin and ring system for spotting. Elements derived from highly expressed mammalian genes (e.g. *Hprt*) can also be used as positional reference markers in key locations on the array (e.g. in each corner of the array) as they ensure a strong signal after hybridisation. Labelling efficiency was monitored by the addition of known quantities (50 and 5 ng) of in vitro transcribed (IVT) RNA from two plant cDNAs to each RNA sample prior to labelling. If labelling efficiencies for two targets are

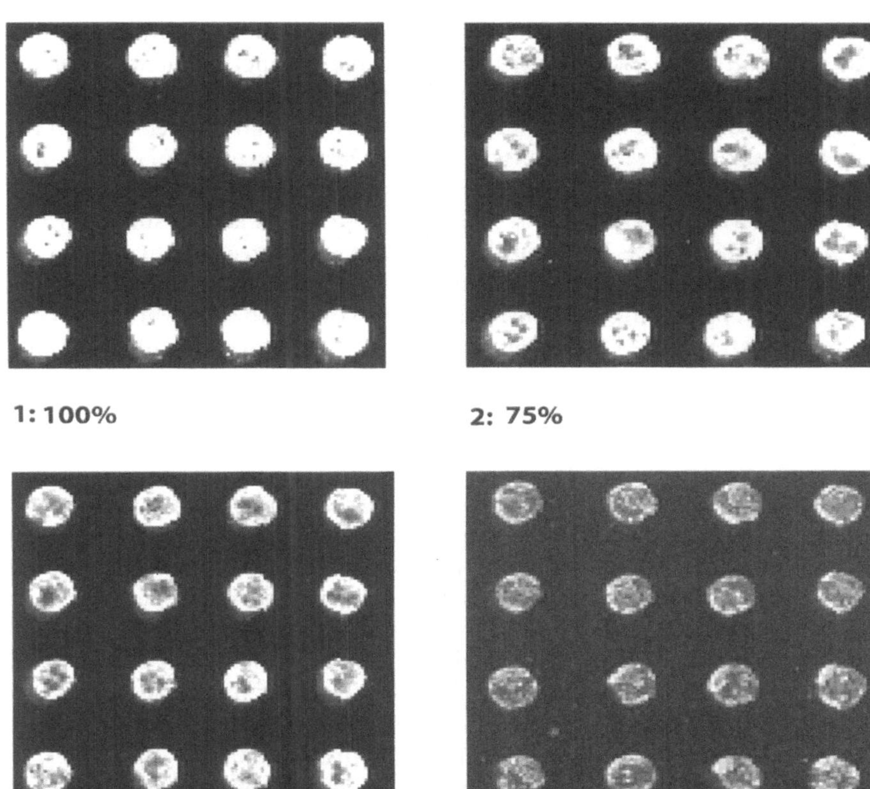

Fig. 2.2. Series of images showing amount of Cy3-labelled PCR product attached to a poly-L-lysine-coated substrate at various stages of a microarraying experiment. **1** Freshly arrayed product; **2** after post-array fixation; **3** after denaturation; **4** after overnight hybridisation in 5×SSC/0.1% SDS at 62 °C. All images were scanned at 70% laser power/70% PMT on a Genetic Microsystems 418 array scanner and Imagene 2.0 (BioDiscovery) software was used to quantify amount of element retained

similar, then similar levels of signal are observed after dual hybridisation. RNA quality was assessed by arraying three distinct elements from the *Gapdh* gene: one from the 5′ UTR region of the transcript, another from the open reading frame and a 3′ UTR element. If degradation affects one of the RNA samples, or if reverse transcription was restricted in one of the labelling reactions, this is detected as uneven signals for the 5′ versus the 3′ *Gapdh* elements from the two targets. Note that this control is only suitable when labelling RNA targets by oligo-dT-primed reverse transcription.

The success of hybridisation and post-hybridisation washes can be monitored by spiking the labelled target mixture with a known quantity of fluorescently labelled probe (50 ng) from two bacterial genes: *amp* and *kan*. The bacterial probes can be labelled by incorporation of fluorescent nucleotides during PCR amplification.

3. Key controls comprise those genes (also known as reference genes) that display a known pattern of expression within the system being studied. It is important to include probes for any genes known to display differential expression within the system you are studying, since it is the performance of these genes on the array that act as important references for the behaviour of anonymous cDNA probes. When all the control genes perform appropriately and a panel of known genes displays expected differential expression patterns, data collected from unknown elements can be used with a high level of certainty. A second reason for including probes for known genes relevant to the biological system of interest is that these can act as useful references when examining the expression data using clustering algorithms. Recent advances in data mining have shown it is possible to cluster genes that show similar expression patterns over a series of experiments; in many cases genes which display similar patterns fall into similar functional pathways (Eisen et al. 1998.) Broadly speaking, housekeeping genes fall into this category because they are expected to display constant expression patterns between any test and reference RNA sample. In practice, however, very few genes fail to exhibit some degree of transcriptional modulation in certain contexts.

3.2.3 Element Preparation

Element preparation can be separated into the following steps: preparation of template for each element, PCR amplification and element purification. The success of each PCR is determined by gel electrophoresis. Optimal probe concentration is in the order of 200–500 ng/μl. Elements arrayed from more dilute solutions generally give poor signal on hybridisation. The most common source of elements in use today are cDNA clones propagated in bacteria. The inserts from each clone must be amplified by PCR and then purified from agents that inhibit DNA attachment to the glass substrate. This PCR amplification is typically done using universal vector-based primers to minimise expenditure on amplimers, though gene-specific primers can be used.

The template for these reactions is typically a small aliquot of bacterial culture or purified plasmid DNA. Our experience with the use of cleared bacterial cultures as a template suggests that successful amplification is achieved in approximately 90% of reactions. Amplifying from purified plasmid DNA resulted in very similar levels of success (greater than 90%) but added a considerable amount of effort to progression from elements to arrays. Plasmid preparation did allow for routine sample sequencing of plasmid templates prior to amplification as a guard against errors with plate orientation and plate selection.

PCR product purification can be performed by a variety of methods ranging from column purification (e.g. Qiagen, Telechem, Millipore) to simple isopropanol precipitation. Great care must be taken at each stage to prevent cross-contamination of probes. The preparation of large numbers of bacteria and plasmids in 96-well format and the subsequent amplification of thousands of PCR products by universal vector primers is prone to cross-contamination. Sample sequencing of plasmid templates, diligent use of negative controls and adherence to good PCR technique are imperative to avoid possible disaster.

3.2.4 Arrays

Arraying can be broken down into the following aspects: (1) choice of a robotic arraying system; (2) choice of a suitable slide chemistry for DNA attachment; (3) choice of an arraying solution; and (4) post-array fixation.

3.2.4.1 Choosing an Arraying System

The last couple of years have seen a vast expansion in the number of robotic arraying units available on the market. There is a clear trend towards faster, higher throughput microarrayers, with the capacity to print a larger number of slides simultaneously. The delivery system aspect of microarraying has also advanced dramatically and there are now a variety of pin, quill, capillary, piezo-electric and ink-jet spotting systems being used to print DNA elements onto substrates. Time should be taken to assess which system is most appropriate to your needs. Array regularity, spot uniformity, speed and efficiency of spotting are the most critical factors that should be considered when deciding on an arraying system.

3.2.4.2
Slide Chemistry

Slide chemistry consists of the way DNA is attached to the non-porous glass substrate and the subsequent inactivation of the substrate post-arraying. When this is done efficiently, DNA elements are successfully bound to the glass surface and then inactivated to prevent labelled cDNA target from binding to the substrate during the hybridisation, resulting in undesirable background. The sensitivity of this technology derives from the ability to detect weak fluorescent signals at a given element on a non-porous substrate in comparison to the very low background surrounding the element.

No matter what arraying system is being used, it is important to determine the best substrate and binding chemistry for a given objective. Some chemistries require the use of modified 5′ amino-linked primers in the PCR amplification used to generate probe. Traditionally, poly-L-lysine- or silane-coated slides have been used. The most popular to date has been poly-L-lysine slides prepared "in house". The advantages of the poly-lysine chemistry are that it requires no DNA modification, it is extremely cheap and, once perfected, it provides a highly consistent performance. The problems associated with home manufacture originate from batch variation in the product, the shelf-life and the poorly understood maturing process that is required between slide coating and microarraying. Commercial poly-L-lysine-coated slides have tended to be substandard for microarray purposes because they were not designed or packaged for this purpose.

More recently, there has been a dramatic increase in the number of commercially available substrates for microarraying. These substrates are generally made using superior glass of uniform thickness and both monolayer (e.g. Corning, Telechem) and branched polymer (e.g. 3D Surmodics) substrates exist. Many of the newer substrates require the use of 5′ amino-modified amplimers for element preparation.

3.2.4.3
Array Solution

As previously mentioned, spot consistency is very important for expression profiling. The nature of spotting can be altered by the solution the DNA is resuspended in. Traditional arraying methods involved the spotting of elements in a salt-buffered solution (2–5×SSC), but the use of surfactants and chaotropic agents (e.g. 1 M NaSCN) contributes to a more even deposition of element onto the glass substrate. We employed DMSO (10–25%) in our spotting solution because it promotes denaturation and reduces the evaporation rate of samples to be printed. The selection of an arraying solution is often determined by the chemistry that is used to attach DNA elements to the glass substrate. Also, it is important to remember that spotting solutions will perform differently depending on the element delivery system.

3.3 Labelled RNA Targets

One of the most challenging aspects of expression profiling experiments is the efficient and consistent labelling of RNA. If the source of RNA is limited, then linear amplification protocols may be required prior to labelling to ensure adequate amounts of starting material (Kacharmina et al. 1999).

3.3.1 Preparing RNA

Differing schools of thought exist on whether total or poly A+ RNA is the best source of RNA template for labelling: each type of RNA may introduce its own particular bias to the experiment. In either case, the RNA must be free of genomic DNA, solvents or other common contaminants retained from simple RNA extractions. The most common cause of problems associated with labelling of RNA targets relates to the quality of the RNA template.

In the case of mammalian tissues that are traditionally known as difficult tissues to extract clean RNA from, muscle for example, caesium banding procedures are recommended. In the case of cell line studies, RNA extraction with guanidinium thiocyanate followed by column purification (e.g. Qiagen *RNAeasy*) is sufficient.

3.3.2 Labelling RNA

Traditionally, RNA templates have been labelled by the incorporation of dye-labelled nucleotides into a first strand synthesis reaction using MLV reverse transcriptase (DeRisi et al. 1996). These reactions are extremely inefficient and as a result require high concentrations of dye-labelled nucleotides during labelling, most of which is discarded. A variety of new protocols have been adapted to circumvent wastage of dye-labelled nucleotides, including recycling of unincorporated label by HPLC (see http://cmgm.stanford.edu/pbrown/mguide/hplc.html).

Recently, the non-covalent attachment of fluorophores to nucleic acid involving a chemical reaction instead of the traditional enzymatic process has become commercially available (http://www.probes.com/media/pis/mp21650.pdf). Also, efforts have been put into the generation of labelling systems allowing cDNA synthesis using modified nucleotides, which incorporate far more efficiently than the traditional Cy-dye-labelled dNTPs. Fluorescent dyes are then ester linked to free amino groups present on the modified nucleotides incorporated into the first strand cDNA(http://

cmgm.stanford.edu/pbrown/protocols/aadutpcouplingprocedure.html). Such labelling methods are substantially cheaper than the traditional labelling protocols given the fact that mono-reactive Cy dyes are one-tenth the cost of Cy-dye-labelled nucleotides.

When RNA samples are limiting, there are several alternatives: protocols that combine RNA target amplification, efficient labelling and signal amplification have also been commercialised (e.g. MicroMax NEN Dupont). Many of these new methodologies are in their early stages of development so it is still unclear how well they compare to the traditional labelling method. Another recent advance has seen the use of dye-labelled-oligo dendrimers to amplify hybridisation signals (Genosphere, http://www.genisphere.com).

A final comment concerning labelling procedures is that good data are dependent on attaining the best signal-to-background ratio possible. Efficient removal of unincorporated products, avoiding evaporation of hybridisation solution while on the microarray and cleanliness of reagents therefore must be a priority.

3.3.3 Hybridisation

Hybridisation conditions for microarrays classically mimic those of other molecular methods. Stringency is controlled by using reagents or parameters common to liquid hybridisation solutions (e.g. SSC, formamide and temperature control). Non-specific background is also prevented by using familiar agents such as SDS and Denhardt's solution. All hybridisation solutions contain large amounts of blocking agents designed to minimise hybridisation due to the presence of repeat sequences or poly A tracts on labelled cDNAs.

The very small volumes used for hybridisation make evaporation a critical factor in experimental failure. Evaporation leads to areas of local background where the hybridisation solution dries onto the surface of the array and cannot be removed by washing. Hybridisation chambers are a common tool used to reduce the chance of evaporation and thus avoid ruining an experiment. The array is hybridised with the labelled target under a coverslip and sealed into a shallow, humidified chamber.

Bubbles need to be avoided when applying coverslips as they lead to local variations in background. The two most common ways of applying the hybridisation solution to the array are to (1) use capillary action to draw the hybridisation solution under the coverslip or (2) deposit the hybridisation solution onto the array and overlay the coverslip. Poor coverslip placement can lead to wicking of the hybridisation solution down the side of the slide and into the chamber.

3.3.4 Scanning and Image Analysis

Microarray readers can be broken into two major classes: confocal laser scanners and CCD (charged couple device) camera-based devices. Each class has its own advantages and disadvantages. CCD systems are often more affordable since the mode of illumination is cheaper and the fluorescent signal can be collected over long periods of time. Such systems also are amenable to use of a wide variety of fluorophores due to the wide excitation spectrum of their illumination sources. This utility comes at a price, however, since these systems traditionally have had reduced signal sensitivity, problems with non-specific signal discrimination, and poor resolution of the collected image.

The confocal laser scanning approach gives excellent sensitivity and high resolution (5–10 μ as standard). In some cases multiple lasers are now used for dual colour detection system and image analysis.

3.3.5 Validation of Leads

In order to guarantee the production of reliable data sets it is best to replicate microarray experiments. It is currently recommended that four sets of data should be generated from each element. When test and reference RNAs are not in limiting amounts, four independent experiments are performed. Alternatively, duplicate experiments are sufficient if each element has been spotted in two positions on the array.

Several studies have used independent methods to accurately validate expression profiles observed with microarrays, using quantitative assays, e.g. quantitative RT-PCR and Northern blotting. Because the majority of our microarray experiments have focused on expression in whole developing mouse organ systems, we chose whole-mount in situ hybridisation for independent validation of our data. Whole-mount in situ hybridisations allowed us to confirm differential expression between organs and also defined cell- or tissue-specific expression in the organ of interest.

4 Conclusions

In this chapter we have tried to highlight some of the issues that are critical for the design and implementation of expression profiling experiments using cDNA microarrays. There are several topics that have not been covered, such as approaches to data normalisation and data mining, but these are dealt with elsewhere in this volume.

It is important to emphasise a flexible approach when using microarrays: utilise those resources and conditions that meet your needs. This technology is developing at such a fast rate that many of the approaches described here will probably be superseded in the coming months and years. It is important that researchers take time to develop protocols and try new approaches, rather than slavishly following published protocols. The protocols included here are meant to serve as a guide, and no more.

In our opinion, some variant of the approach described here is likely to become as important to basic biological research, as well as clinical research, as PCR methodology has been in the last decade. Whether developmental biology or toxicology, most areas of research will benefit enormously from the insights offered by expression profiling of complete genome sets in the near future.

5 Protocols

The protocols described here have been used by us routinely for expression analysis in the developing mouse embryo (Grimmond et al. 2000). They were employed in conjunction with hardware from GMS/Affymetrix and software from BioDiscovery. They are derivations of protocols of Schena and Davis, the lab of P. Brown (Stanford) and Hegde et al. (2000). The original protocols can be found at http://arrayit.com/dna-microarray-protocols and http://cmgm.stanford.edu/pbrown/protocols/index.html.

5.1 Poly-L-Lysine-Coated Substrate Preparation

We follow the protocol from the Brown lab web site(http://cmgm.stanford.edu/pbrown/protocols/1_slides.html).

5.2 Element Preparation

5.2.1 Plasmid Preparation

Elements are normally prepared using an alkaline lysis method that is performed in 96-well format using multi-channel pipettes. Generally, the DNA is of good enough quality for sequencing.

1. Aliquot inoculate 1.1 ml of LB broth + ampicillin 50 µg/ml into 2.2-ml-deep 96-well plates. (Plates with square wells and round bottoms are best.)

Cover the plate with sealable lid and shake at 250 rpm (minimum) overnight at 37 °C.

2. Spin down bacterial pellet by spinning plate at 2,000 *g* for 15 min, 4 °C.
3. Tip off supernatant and pat dry on towel.
4. Resuspend bacterial pellet in 100 µl of buffer P1 (10 mM Tris pH 8.0, 1 mM ETDA pH 8.0, 100 µg/ml RNase A and RNase T1. Leave for 5 min at ambient temperature.
5. Add 100 µl of buffer P2 (0.2% SDS, 0.2 M NaOH). Leave on ice for 5 min.
6. Add 100 µl of 3 M Na acetate pH 4.8 and precipitate at –20 °C for 30 min.
7. Spin plate at 3,000 *g* for 60 min at 4 °C and transfer 125 µl to a new 96-well plate. Add 125 µl of isopropanol and precipitate at –20 °C for at least 1 h.
8. Precipitate plasmid DNAs by spinning plate at 3,000 *g* for 60 min at 4 °C. Tip out supernatant. Wash with 500 µl of 70% ethanol. Spin at 3,000 *g* for 10 min at 4 °C. Leave plate for 60 s to dry on bench.
9. Resuspend in 50 µl of TE and run 5 µl on a gel to check the preparation. Dilute 10 µl down to 2 ng/µl for generating PCR products.

5.2.2
PCR Amplification

1. PCR reactions are set up as follows:

Reagent	Per reaction	Per 4×96 reactions
10×PCR buffer	10	4.8 ml
25 mM $MgCl_2$	7	4.0 ml
100 mM NH_2-M13F*	0.3	140 µl
100 mM NH_2-M13R*	0.3	140 µl
5 U/µl Taq (AB Biotech)	0.8	320 µl
H_2O to		48 ml

The primers (*) have a 5′ NH_2 linker group with a C_6 spacer. The primer sequences (17) are:

NH_2-M13F NH_2-C6-GTT TTC CCA GTC ACG AC
NH_2-M13R NH_2-C6-ACA GGA AAC AGC TAT GAC

2. Dispense bulk mix into a standard profile polypropylene 96-well plate (Thermofast, AB Biotech). Low profile plates are not suitable since they cannot hold the volumes required for isopropanol precipitation of the PCR products.
3. Add 5 µl of plasmid DNA to each reaction.
4. PCR reactions are carried out in an MJ research tetrad PCR machine under the following conditions: (temperature of block is calculated, plate option is selected, volume of reaction selected is 100 µl in the program setup) 1×94 °C for 180 s, 35×(48 °C for 20 s, 72 °C for 180 s, 94 °C for 40 s).
5. Run 5 µl of the PCR reaction on a 96-lane gel to determine success of the PCR amplification. Record failures and then re-amplify corresponding samples.

5.2.3
Product Purification

1. Immediately after electrophoresis, add 10 µl of 3 M Na acetate (pH 5.2) and 100 µl isopropanol. Seal the plate and precipitate at -20 °C overnight. Elements are recovered by spinning at a minimum of 3,000 rpm for 60 min at 4 °C. Invert plates onto an absorbent towel and gently tap dry. Pellets are often just visible.
2. Wash pellets with 500 µl of 70% ethanol. Spin at 3,000 rpm for 10 min at 4 °C and blot plates dry on fresh towel.
3. Leave to air dry for 60 min to overnight. Seal plates and store at –20 °C until ready to array.
4. Prior to arraying, resuspend pellets in the desired array solution.

5.3
Post-Array Fixation

Once again, this protocol was adapted from the Brown web site (http://cmgm.stanford.edu/pbrown/protocols/3_post_process.html).

5.4
RNA Target Labelling

We have tried all sorts of variations on the theme of reverse transcriptase (RT) incorporation of Cy-dyes into first strand cDNA. We have performed this on mRNA prepared by a variety of kits (Invitrogen, Pharmacia and Qiagen), and when it did work, the Qiagen kit was probably the best. RNA preparation is a *critical* step for ensuring good results. In our hands, results are entirely dependent on RNA quality. Anything less than perfect and we generally saw no signal.

More recently, we have begun to use total RNA. We have used the Qiagen RNeasy kit and seen an improvement in the robustness of our labelling experiments. Note: this works well on embryonic tissues that lyse very easily and contain little connective tissue at the stages at which we work. The use of adult tissues may require far more diligent preparation (e.g. homogenisation in GITC and CsCl banding).

RNA is prepared using RNeasy protocol. A Qiashredder is incorporated to minimise gDNA contamination. We also check the purified RNA with primers from the 3′ UTR of *Gapdh* in a control RT-PCR experiment to confirm the absence of genomic DNA. If genomic DNA is present, the RNA should be treated with RNase-free DNase for 30 min at 37 °C followed by inactivation of the DNase at 70 °C for 10 min. RNA is made just prior to labelling when possible, or stored at –70 °C if necessary.

The following protocol has been used routinely with 20 μg of total RNA. Up to 100 μg of total can be used; 1–5 μg of mRNA can also be used.

1. Mix the following:

RNA (10 μg/μl)	2 μl
Oligo-dT (T_{18-25}) (2 μg/μl)*	2 μl
H_2O	20 μl

2. Heat to 70 °C for 10 min. Cool to 37 °C by putting into preset hot block for 10 min.
3. Prepare the following:

5×RT buffer (BRL)	10 μl
DTT	5 μl
25.0 mM dATP, dTTP, dGTP	1 μl
2.5 mM dCTP	2 μl
1.0 mM Cy3 or Cy5 dCTP	4 μl
RNA-Guard	1.5 μl

4. Add 23 μl of the above mix to the RNA + oligo dT mix. Spin down with 10-s pulse in the centrifuge and replace at 37 °C for 1 min. Add 2 μl of RTase (Superscript II, BRL Life Technologies). Incubate for 90 min at 37 °C.
5. Add 1 μl of 0.5 M EDTA and 2 μl of 2 M NaOH. Heat to 65 °C for 10 min. Chill on ice for 1 min and add 4 μl 1 M HCl and 4 μl 1 M Tris pH 8.0.
6. Remove unincorporated dye-labelled nucleotides by ethanol precipitation. Add 10 μl of 1 μg/μl Cot1 DNA (Life Technologies), 7 μl of 3 M Na acetate (pH 4.8) to the reaction. Mix well. Add 150 μl of ethanol and chill at –20 °C for 30 min. Spin down the pellet (14,000 rpm, 4 °C, 20 min) and wash pellet well with 500 μl of 70% ethanol. Allow pellet to air dry but do not over dry.
7. Resuspend both pellets in the same 50 μl of microarray hybridisation solution (4×SSC/0.2% SDS/50% formamide).
8. Add 1 μl of hybridisation standards and heat to 80 °C for 10 min to denature probe. Cool on ice and add 1 μl 10 mg/μl poly dA (Pharmacia). Incubate the labelled target at 42 °C for at least 1 h prior to placing on array. This competes out repeat sequences and poly T tracts.
9. Cool hybridisation mix on ice for 1 min (this prevents a halo from hot hybridisation solution drying on the slide). Spin for 1 min at maximum speed to collect condensation and to precipitate any possible particulate material.
10. Place hybridisation solution onto array, avoiding bubbles! This is best done by carefully laying the coverslip onto slide with fine forceps. Alternatively, slowly pipette the hybridisation solution next to a coverslip placed over a dry slide. Capillary action draws the hybridisation solution under the coverslip. Applying hybridisation solution is worth practising, especially with the large 60-mm coverslips.
11. Place slide into hybridisation chamber. Add 10 μl of hybridisation mix into the wells at each end of the chamber. Tighten lid and submerge in 45 °C water bath for 14–24 h.

12. After hybridisation, dry chamber, place slide with coverslip into slide holder and plunge into 1 l of 0.2×SCC, 0.05% SDS at ambient temperature until coverslip falls off. Rock for 3 min. Rapidly transfer slide holder into a second beaker of 1 l 0.2×SSC and rock for a further 3 min.
13. Carry beaker plus slides to centrifuge set at ambient temperature. Slides can be conveniently spun in a 50-ml plastic V-bottomed tube if racks are not available. Remove slide with forceps and spin at 500 rpm for 3 min. If possible, put the "handled" end of slide into the carrier so that is spins on the outside during centrifugation. This prevents contaminants getting onto the array.

5.5 Controls

5.5.1 Sensitivity Controls

We use several sets of controls: the first set is a series of pre-labelled PCR products that are added at known concentrations to each hybridisation. These labelled products allow the efficiency of hybridisation to be gauged.

Amp PCR product	50 ng/hyb
Kan PCR product	5 ng/hyb

1. Products are labelled by incorporating Cy dyes during amplification.
2. For a 25-µl reaction:

10×PCR buffer (plus $MgCl_2$)	2.5 µl
Low CTP mix (2 mM A, G, TTP, 0.4 mM dCTP)	2.5 µl
Cy3 or Cy5 CTP (1 mM)	2.0 µl
10 µM oligos	2.0 µl
Taq (5 U/µl)	0.5 µl
H_2O to	25 µl

3. Prepare a Cy3- and Cy5-labelled product for each control gene. Use standard amplification procedures. Remove unincorporated Cy-CTP by Qiaquick PCR product purification column.
4. Dilute Cy3 and Cy5 products for each control gene to the same concentration. Make up stock tube that contains labelled gene products at the following concentrations: Cy3 and Cy5 *Kan* products, 50 ng/µl; Cy3 and Cy5 *Amp* products, 5 ng/µl; and Cy3 and Cy5 Plant1 gene products, 500 pg/µl.

5.5.2 Labelling Controls

To our target RNA we add in vitro RNA generated from a series of plant genes known to contain large poly A tails. The non-mammalian RNA is added prior

to each reaction to determine the efficiency of the RT step. RNA is made using the Ambion RNA megascript kit. RNA samples are diluted out to the following concentrations and added to hybridisations:
Plant gene 1 in vitro RNA:50 ng/hyb
Plant gene 2 in vitro RNA: 5 ng/hyb

5.5.3 RNA Integrity Control

We also print three *Gapdh* PCR products onto each array. These products are derived from the 5′ UTR, ORF and 3′ UTR regions of the *Gapdh* transcript. If RNA is degraded in one of the samples or if the RT labelling has generated truncated products, all three elements will not give a strong co-expressed signal.

References

Alizadeh AA, Eisen MB, Davis RE, Ma C, Lossos IS, Rosenwald A, Boldrick JC, Sabet H, Tran T, Yu X, Powell JI, Yang L, Marti GE, Moore T, Hudson J Jr, Lu L, Lewis DB, Tibshirani R, Sherlock G, Chan WC, Greiner TC, Weisenburger DD, Armitage JO, Warnke R, Staudt LM (2000) Distinct types of diffuse large B-cell lymphoma identified by gene expression profiling. Nature 403(6769):503–511

Bonaldo MF, Lennon G, Soares MB (1996) Normalization and subtraction: two approaches to facilitate gene discovery. Genome Res 6(9):791–806

DeRisi J, Penland L, Brown PO, Bittner ML, Meltzer PS, Ray M, Chen Y, Su YA, Trent JM (1996) Use of a cDNA microarray to analyse gene expression patterns in human cancer. Nat Genet 14(4):457–460

DeRisi JL, Iyer VR, Brown PO (1997) Exploring the metabolic and genetic control of gene expression on a genomic scale. Science 278:680–686

Dixon AK, Richardson PJ, Lee K, Carter NP, Freeman TC (1998) Expression profiling of single cells using 3 prime end amplification (TPEA) PCR. Nucleic Acids Res 26(19):4426–4431

Eisen MB, Spellman PT, Brown PO, Botstein D (1998) Cluster analysis and display of genome-wide expression patterns. Proc Natl Acad Sci USA 95(25):14863–14868

Grimmond S, Van Hateren N, Siggers P, Arkell R, Larder R, Soares MB, Bonaldo M, Smith L, Tymowska-Lalanne Z, Wells C, Greenfield A (2000) Sexually dimorphic expression of protease nexin-1 and Vanin-1 in the developing mouse gonad prior to overt differentiation suggests a role in mammalian sexual development Hum Mol Genet 9(10):1553–1560

Hegde P, Qi R, Abernathy K, Gay C, Dharap S, Gaspard R, Earle-Hughes J, Snesrud E, Lee N, Quackenbush J (2000) A concise guide to cDNA microarray Anal Biotech 29(3):548–562

Kacharmina JE, Crino PB, Eberwine J (1999) Preparation of cDNA from single cells and subcellular regions. Methods Enzymol 303:3–18

Luo L, Salunga RC, Guo H, Bittner A, Joy KC, Galindo JE, Xiao H, Rogers KE, Wan JS, Jackson MR, Erlander MG (1999) Gene expression profiles of laser-captured adjacent neuronal subtypes. Nat Med 5(1):117–122

Schena M, Shalon D, Davis RW, Brown PO (1995) Quantitative monitoring of gene expression patterns with a complementary DNA microarray. Science 270:467–470

Schena M, Shalon D, Heller R, Chai A, Brown PO, Davis RW (1996) Parallel human genome analysis: microarray-based expression monitoring of 1000 genes. Proc Natl Acad Sci USA 93:10614–10619

Soares MB, Bonaldo MF, Jelene P, Su L, Lawton L, Efstratiadis A (1994) Construction and characterization of a normalized cDNA library. Proc Natl Acad Sci USA 91(20):9228–9232

CHAPTER 3

cDNA Microarrays on Nylon Membranes with Enzyme Colorimetric Detection

KONAN PECK[1] and YUH-PYNG SHER[1]

[1] Institute of Biomedical Sciences, Academia Sinica, 128 Sec. 2, Yen Chiu Yuan Road, Taipei, Taiwan 115, R.O.C.

1 Introduction

Enzyme colorimetric detection is an alternative to fluorescence detection that can be used when simultaneously measuring multiple parameters is required (Tijssen 1985; Thorpe and Kerr 1994). Colorimetric detection methods combined with the use of microarrays (microarray/CD) (Chen et al. 1998) make the already powerful microarray method (Schena et al. 1995; DeRisi et al. 1997; Iyer et al. 1999) even more useful to researchers. In this chapter, we describe how to use the enzyme colorimetric detection method in a microarray format using nylon membranes to simultaneously measure the expression level of multiple genes.

2 Fabrication of Microarrays on Nylon Membranes

2.1 Construction of Non-redundant Gene Sets

For cDNA microarrays, immobilized targets are typically collected from cDNA libraries, subtracted and normalized according to the intended applications. For example, libraries constructed by subtracting cDNA from normal and diseased cell lines are usually used for enriching differentially expressed genes in cancer studies. Several PCR-based methods, such as representational difference analysis (RDA) (Hubank and Schatz 1994; Braun et al. 1995; Chu and Paul 1997), are commonly used to construct subtracted cDNA libraries. The subtraction process leads to enrichment of differentially expressed genes by approximately ten-fold or so, and the redundancy of cDNA clones representing the same gene tends to be high.

Expressed sequence tags (ESTs), the short sequence segments produced by partial sequencing of cDNA clones, have been generated in large numbers and are available in public EST databases (Boguski and Schuler 1995). The ESTs of several organisms are available in the dbEST database of the National Center for Biotechnology Information (NCBI). EST clones were generated from cDNA libraries with or without subtraction. Primers complementary to the DNA sequences flanking the insertion site on the vector are used to sequence the inserts from both the 5′ and the 3′ ends. About 200–500 nucleotides are sequenced at each end. One of the major contributors to the information contained in dbEST, the IMAGE consortium (Lennon et al. 1996), has arrayed more than 3 million cDNA clones and deposited more than 2 million EST sequences in the dbEST database.

Several groups have analyzed the sequences contained in the dbEST database and the known gene sequences in the GenBank database (Benson et al. 1977) using various clustering algorithms to construct gene indices of

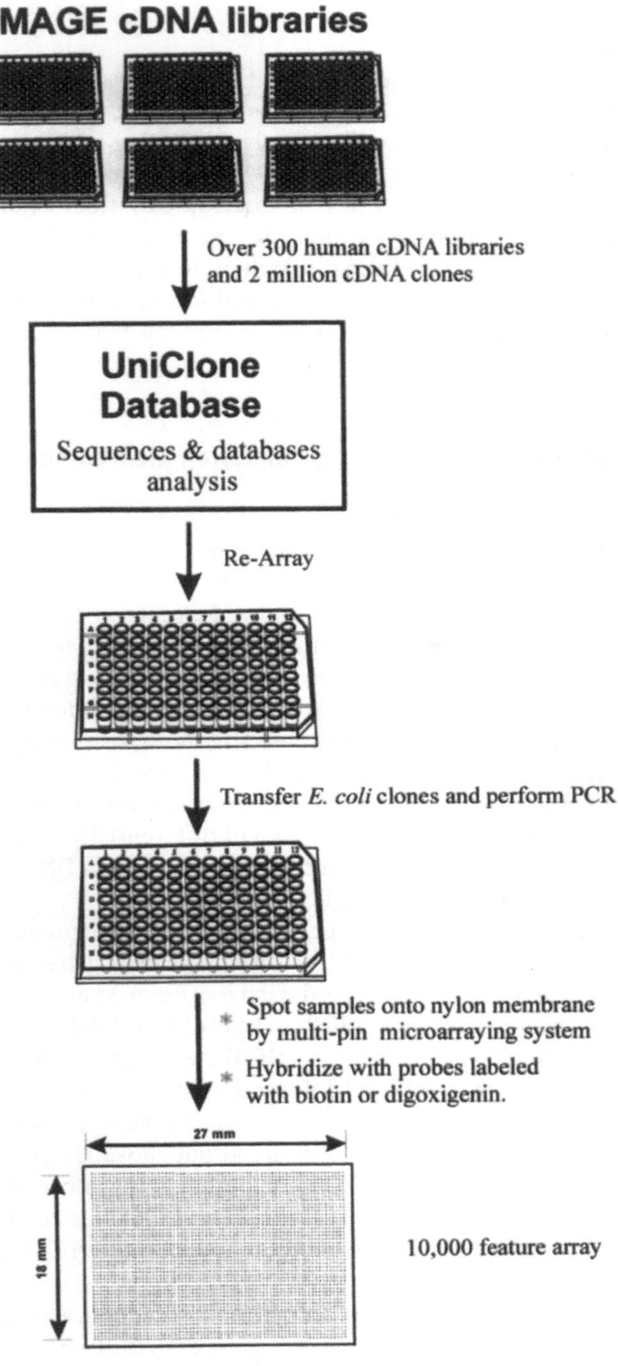

Fig. 3.1. Array fabrication based on IMAGE cDNA libraries and non-redundant cDNA clones

organisms. For example, the NCBI archives putative gene clusters for several organisms in the UniGene clustering (Schuler et al. 1996) and the Institute of Genome Research (TIGR) have constructed the TIGR Gene Indices for human, mouse, rat, and other organisms (http://www.tigr.org/tdb/tgi.shtml). The IMAGE consortium's IMAGEne database (Cariaso et al. 1999) contains full-length IMAGE cDNA clones, as well as clustered cDNA clones to represent putative genes. As of May 2000, the UniGene human EST clustering (Build #112) contained 89,632 putative gene clusters in about 1.5 million human sequences.

For genome-wide expression analysis of human or mouse genes, it is much simpler to use cDNA clones whose sequences and related information are already in the public domain than to construct new and separate cDNA libraries. Based on the UniGene and IMAGE human cDNA libraries, we constructed a set of non-redundant human cDNA clones to represent the putative human genes in UniGene clusters, with the goal of having one representative clone for each gene. We dubbed the non-redundant clone library the "UniClone library". This information of the library is available on an anonymous ftp server, ftp://genestamp.ibms.sinica.edu.tw.

Figure 3.1 depicts the flowchart of how we derived the re-arrayed UniClone cDNA clone set. For ease of preparation, clones with the same cloning vectors were grouped together in the same plates. To check the quality of clones and PCR amplifications, each PCR amplification product was separated using agarose gel electrophoresis (protocol 1). When gels indicated that clones were cross-contaminated with other clones or that PCR failed, the clones were discarded and alternative clones in the same putative gene cluster were substituted.

2.2 The Automatic Arraying Machine

Among all arraying methods, mechanical spotting is currently the most widely used method of cDNA microarray fabrication. The mechanical spotting method employs a print head to transfer reagents onto solid substrates, such as glass slides or nylon membranes. Solid pins or pins with cut-slits are used. Solid pins hold reagents at the tip by surface tension, and the amount of reagent held depends on the diameter of the pin at the tip. A solid pin makes only one reagent deposit per round and goes back-and-forth between the reagent reservoir and the solid substrate. Pins with cut-slits hold reagents by capillary action and are able to continuously deposit reagent drops onto solid substrates. The amount of reagent deposited on the substrate depends on the width of the slit. Obviously, the spotting rate is higher for pins with cut-slit. Based on our experiences and the specifications of commercial arrayers, the coefficient of variation (CV) of spot size is about 7% for solid pins and about 10–15% for pins with cut-slits.

For genome-wide expression studies, tens of thousands of gene targets are usually employed. For such large-scale operations, an arrayer with a plate stacker is required. In our lab, we developed two high-capacity arraying machines with plate stackers based on two different designs. One of the machines is equipped with a rotary plate stacker and holds approximately 31,000 samples. The other arrayer utilizes linear motors to increase the motion speed and has towers of plate stackers capable of holding more than 100,000 samples at a time. Both systems are capable of performing walk-away operations and can print arrays with a spot density exceeding 6,000 spots/cm^2.

2.3 Nylon Membranes

Nylon membranes have been used as a blotting substrate for more than a decade. Compared with glass slides, nylon membranes have a larger surface area and can accommodate more DNA molecules. PCR products can be directly deposited on nylon membranes without prior removal of PCR primers. DNA molecules can be readily immobilized on positively charged nylon membranes with or without UV cross-linking. The binding of DNA molecules to nylon membranes is strong enough to withstand the stripping process, and arrays can be re-probed several times without degradation of signal-to-noise ratio.

3 Enzyme Colorimetric Detection

3.1 Multi-color Enzyme Colorimetric Detection

The principles of enzyme immunoassay are well established and the technique is widely used to detect proteins or antigen-antibody complexes in Western blots, ELISA, or dot blots. Enzymes such as alkaline phosphatase, horseradish peroxidase, and β-galactosidase are traditionally used in protein studies to yield chromogens for quantitative measurement. Different combinations of enzyme/substrate pairs generate different chromogens.

Lee et al. (1988) first demonstrated multi-color detection using two colors in Western blots by using alkaline phosphatase and horseradish peroxidase with sequential incubation of enzyme-antibodies. To simplify the process for microarray applications, we developed an accelerated approach by simultaneously incubating various enzyme-antibody conjugates with different hapten-labeled DNA fragments (protocols 2 and 3). While single antibody incubation requires only about 30 min, simultaneous incubation of multiple antibodies requires a significantly longer period. Generally, good results can be achieved with a 2-h incubation time by adding 4% polyethylene glycol 8000 to the antibody mixture (Hellsing and Richter 1974).

In principle, the number of available color-forming enzymes determines the number of distinctive colors in the enzyme colorimetric reactions. However, because different enzyme/substrate reactions yield different detection sensitivities, and because not every enzyme can be easily conjugated onto antibody molecules, only a limited number of enzyme/antibody conjugates are commercially available. These enzyme/antibody conjugates yield different chromogens with different substrates. In multi-color enzyme colorimetric detection, careful selection of enzyme/substrate pairs and the order of color development reactions are important to generating distinctive colors because one enzyme may be quenched by the substrates of other enzymes.

3.2 Color-Forming Enzymes

In two-color microarray systems, the two color-forming enzymes, β-galactosidase and alkaline phosphatase, have different reaction mechanisms and can be used with minimal cross-interference. β-Galactosidase catalyzes hydrolysis of the non-reducing β-D-galactose residues in β-D-galactosides to produce β-D-galactose and alcohol (Kerr et al. 1994). The commonly used substrate of ß-galactosidase in enzyme immunoassay is X-gal (5-bromo-4-chloro-3-indolyl-β-D-galactopyranoside), which yields a blue-colored product after reacting with the β-galactosidase enzyme. The major substrate of alkaline phosphatase is orthophosphoric monoester, and the enzyme acts to hydrolyze the phosphate ester bond. Many diazonium salts can be used to produce different chromogens (Gossrau 1978). The Fast Red TR substrate is reportedly not as sensitive as another commonly used substrate, BCIP (5-bromo-4-chloro-3-indolyl phosphate)/NBT (nitroblue tetrazolium) combination. However, this BCIP/NBT substrate pair produces a purple precipitate. In two-color microarray applications, the red color of Fast Red TR salt (Escribano et al. 1984) is a more appropriate choice.

4 Characteristics of the Microarray/CD System

4.1 Sensitivity

Several formats of DNA arrays are available for gene expression measurements. Although the formats vary in physical dimensions and utilize different detection methods, the detection limits in terms of the number of sample molecules are of the same order of magnitude, i.e., about 10^7 molecules (Bertucci et al. 1999), meaning that in order to detect gene expression as low as one transcript per cell, 10^7 cells are needed. These requirements exclude the use of micro-dissected tissue specimens, since these specimens usually contain only a few hundred cells.

In principle, among the radioactive, fluorescent, and enzyme colorimetric detection methods, the radioactive method should be the most sensitive and the enzyme colorimetric method the least sensitive. However, macroarrays with radioactive detection have larger sizes and require greater sample volumes, which makes the detection limit of this method more or less of the same order of magnitude as the others (Nguyen et al. 1995; Bernard et al. 1996). Signal intensity is dependent on the number of target molecules immobilized on the solid substrate (Bertucci et al. 1999), and porous nylon membranes have a greater surface area to accommodate more target molecules than glass slides. Compared with laser-induced fluorescence detection, the relative insensitivity of the enzyme colorimetric detection method is compensated for by the greater number of target molecules immobilized on the nylon membranes so that the detection limits of the two methods are about the same.

4.2 Throughput

The enzyme colorimetric detection method requires antibody/hapten coupling and color development time, which takes about 3 h longer than the time required for laser-induced fluorescence detection. After color development, microarray/CD images can be digitized with either a standard flatbed digital scanner or a drum scanner commonly used in the pre-press industry. Flatbed scanners are readily available in retail computer stores; alternatively, services for a fee are usually available for image digitization in print shops. For high-density arrays with spot diameters of 100 μm or less, a flatbed scanner with 3,000 dots-per-inch (dpi) optical resolution is recommended. Most drum scanners have 10,000 dpi resolution. A flatbed scanner digitizes a letter size area, which accommodates around 100 arrays each with a size of 1.8×2.7 cm, in less than 10 min. On the other hand, a confocal laser-induced fluorescence scanner takes 3–5 min to digitize an array. The fluorescence scanning speed is based on the photon burst rate of the fluorophore, which is determined by the power density of the laser excitation and the emission lifetime of the fluorophore (Mathies et al. 1990). If the signal sampling time is less than optimal, the detector collects fewer photons, resulting in lower detection sensitivity; hence the digitization time in confocal laser-induced fluorescence detection is a compromise between scanning speed and detection sensitivity. For small numbers of arrays, fluorescence detection has higher throughput. In clinical settings or for applications that require processing hundreds of arrays per day, enzyme colorimetry has a higher throughput if hybridization and color development are automated and done in parallel.

4.3 Dynamic Range

The different detection methods have widely different dynamic ranges. The extent of the range is largely limited by the detection devices employed rather than by the physical principles of the individual detection methods. For example, the dynamic range of radioactive detection can reach five orders of magnitude if an imaging plate is used as the detection device. On the other hand, the dynamic range is rather poor if X-ray film is the detection device. The same conditions are true for laser-induced fluorescence detection. If the detectors are operated in photon counting mode using photomultiplier tubes, the dynamic range can reach 5 orders of magnitude. However, if photodiodes or charge-coupled devices are used as the detectors, the dynamic range is usually less than what can be achieved by photon counting.

The dynamic range of enzyme colorimetric detection is determined by the type of imaging device used and is usually between 0.1 and 4.0 optical density (OD). Although densitometry or absorbance detection using spectrophotometers can have detection ranges from 0.001–4.0 OD, the digital scanners employed in microarray/CD are not designed to have such a wide detection range. Flatbed scanners were originally designed to digitize photographs or documents for which such high sensitivity is not required; hence enzyme colorimetric detection has a much narrower dynamic range. This narrow dynamic range allows for a smaller measurement window than the other detection methods; hence enzyme colorimetric detection is mainly useful as a semi-quantitative method to be used for applications that do not require an extended dynamic range of measurement, such as the identification of differentially expressed genes. SAGE measurements (Zhang et al. 1997) indicate that 86% of all genes are expressed at less than five transcript copies per cell, and 99.8% of these genes have transcript copy number below 50 copies per cell; hence a semi-quantitative method without extended dynamic range may be considered appropriate.

For any microarray system, the minimum detectable differential expression ratio is determined by the cumulative systematic errors of the various steps of the process. For arrays fabricated using solid pins and with single-color enzyme colorimetric detection, signals for the same gene among arrays on different membranes have an approximate 7% coefficient of variation (CV) after normalization with external controls. For two-color detection, the discrimination limit is 1.4-fold in differential expression ratio. These ratios make twofold differentially expressed genes readily detectable using colorimetric methods. At the same time, the differential expression ratio is skewed at the high end due to the methods' limited dynamic range. In practice, enzyme colorimetric methods identify as many differentially expressed genes as laser-induced fluorescence detection; the current bottleneck is in data processing and analysis.

4.4 Sensitivity Enhancement

Due to the large amount of RNA required per hybridization, most applications of cDNA microarrays are currently limited to the use of RNA derived from cultured cells. Several research groups are currently testing various strategies to improve the method's sensitivity. Our strategies for improving sensitivity have involved using signal amplification based on a modified catalyzed reporter deposition (CARD) method (Bobrow et al. 1989, 1991).

The CARD amplification method uses horseradish peroxidase (HRP) as the analyte-dependent reporter enzyme (Zaitsu and Ohkura 1980). Complex DNA probes are labeled with biotin to hybridize to target molecules on the nylon membrane. After hybridization, streptavidin conjugated with HRP is applied to detect the hybrid. In the presence of hydrogen peroxide, HRP reacts with the phenolic part of a biotin-tyramide compound to produce a quinone-like structure bearing a free radical on the C2 group of the tyramide. The activated biotin-tyramide then covalently binds to the electron-rich amino acid residues, such as tyrosine or tryptophan, of protein molecules absorbed on the nylon membrane. Since the free radical is short lived, the solid phase reaction only occurs in the location where it is generated.

Using CARD amplification (protocol 4), we were able to improve the detection limit by 60-fold compared with the regular colorimetric methods. We verified that this modified CARD method yields the same gene expression pattern as the more standard methods by hybridizing 2 μg (regular) and 35 ng (CARD) of biotin16-dUTP labeled poly-A + RNA to two pieces of nylon membranes containing PCR product of 576 EST clones. Regular and CARD-amplified colorimetric detection methods were then applied to the two arrays, respectively, and similar expression patterns were obtained (Fig. 3.2).

In addition to working to improve microarray/CD signal amplification, researchers are also utilizing sample amplification methods (Van Gelder et al. 1990; Eberwine et al. 1992) in microarray/CD to enrich the amount of RNA molecules used for hybridization. These methods allow micro-dissected tissue specimens to be used. Briefly, total RNA is primed with a synthetic oligonucleotide containing the T7 promoter sequence to yield the first strand cDNA. The second strand cDNA synthesis process is then applied to yield double-stranded cDNA. After double-stranded cDNA synthesis, T7 RNA polymerase is used to generate amplified antisense RNA (aRNA). The yield of one round of amplification is typically around 10–20 μg aRNA from 1–2 μg total RNA (Salunga et al. 1999). A second round of amplification results in $>10^6$-fold amplification from the original material.

In our lab, we combine one round of RNA amplification and one round of signal amplification (CARD) to improve detection sensitivity. Figure 3.2 shows a comparison of the expression patterns of 576 genes using three methods: (1) regular method (no amplification) using 2 μg mRNA; (2) CARD method

Fig. 3.2. Gene expression patterns of three different methods. **A** Regular method using 2 μg labeled mRNA. **B** CARD method using 35 ng labeled mRNA. **C** aRNA combined CARD method using 10 ng labeled total RNA

Regular

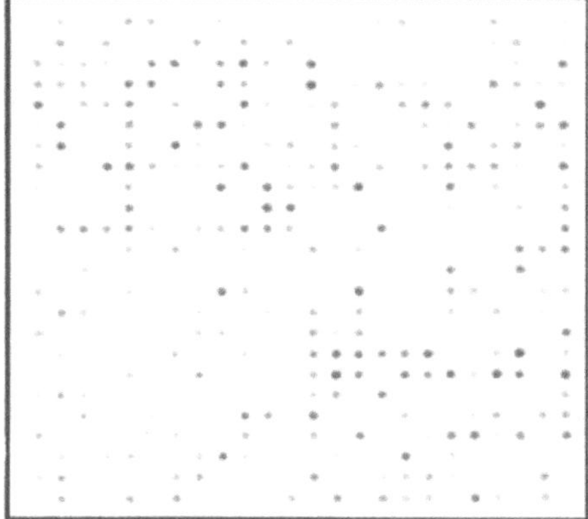

Card

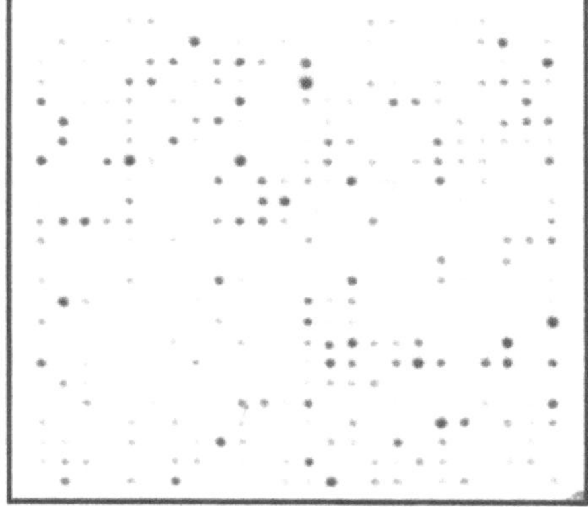

RNA amplification

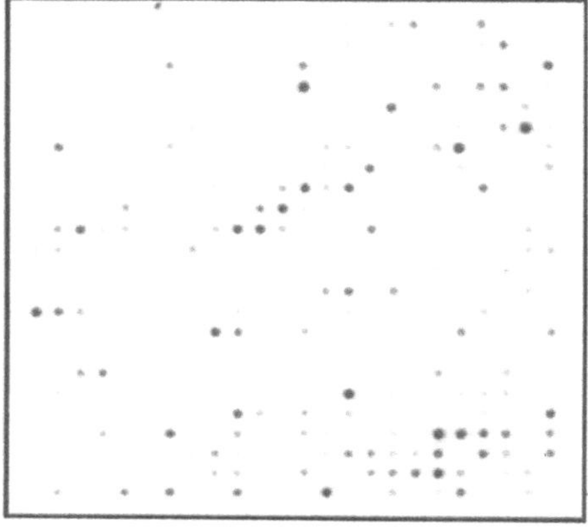

(signal amplification) using 35 ng mRNA; and (3) aRNA with CARD method (RNA and signal amplification) using 10 ng total RNA. The first two methods elicit very similar expression patterns, while the expression pattern of amplified RNA with CARD is slightly different from the other two methods. The deviation may be due to the biased amplification efficiency for RNA molecules of various sequences.

5 Data Processing in the Microarray/CD System

The algorithm required to analyze a two-color image obtained by enzyme colorimetric detection is different from the algorithm used for pseudo-color-encoded images obtained by laser-induced fluorescence detection. cDNA microarrays (Fig. 3.3) using colorimetric detection yield images with a light background and true colors composed of three primary colors, cyan, magenta, and yellow (C, M, Y), rather than a dark background and pseudo-color images as in fluorescence detection. Genes commonly expressed in both normal and diseased (or induced and non-induced) states exhibit various intensities of purple, while differentially expressed genes exhibit red or blue. Figure 3.4 shows a microarray/CD flowchart for imaging and data analysis. To identify differentially expressed genes, the color of each spot in an array is separated into three primary colors and the composition is represented in a 3-D space of CMY. The biotin/digoxigenin ratio of each gene is then calculated based on the CMY composition of a set of control genes labeled with known biotin/digoxigenin ratios. Subsequent data analysis methods are otherwise the same as for other microarray systems.

Compared with laser-induced fluorescence detection, the major advantage of enzyme colorimetric detection is its accessibility: a standard consumer level flatbed scanner can be employed for detection, and no expensive optical instruments are required. However, microarray/CD, while convenient, also limits the linear response of detection and compresses a large expression difference to a smaller than actual number. Consequently, until colorimetry scanners with higher dynamic range are available, the enzyme colorimetric detection method is not suitable for measuring the exact ratio of differential expression. The detection method is semi-quantitative and suitable only for applications that screen for differentially expressed genes beyond a set threshold of expression difference. For example, the method discriminates between differentially expressed genes with expression ratios range from twofold to tenfold, but is unable to distinguish a 100-fold differentially expressed gene from a 200-fold differentially expressed gene.

The number of differentially expressed genes varies from one cell system to another. For example, with enzyme colorimetric detection, cancer cells treated with or without anti-cancer drugs can result in 0.6–1% of genes having at least

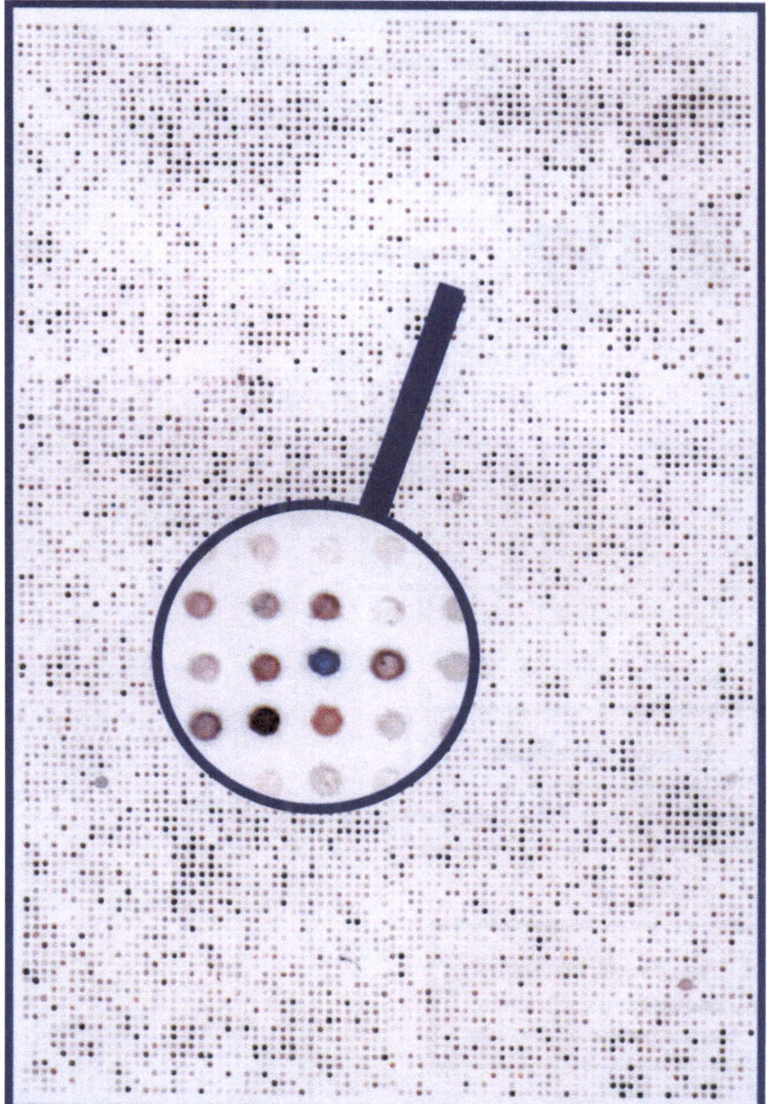

Fig. 3.3. A two-color cDNA microarray image using colorimetric detection method. Microarray image is the expression pattern of KB cells treated with and without camptothecin. *Blue* and *red spots* represent two extremes of camptothecin's effect on gene expression. *Blue spots* represent genes whose expression was repressed by camptothecin and *red spots* represent genes whose expression was induced by anti-cancer drug

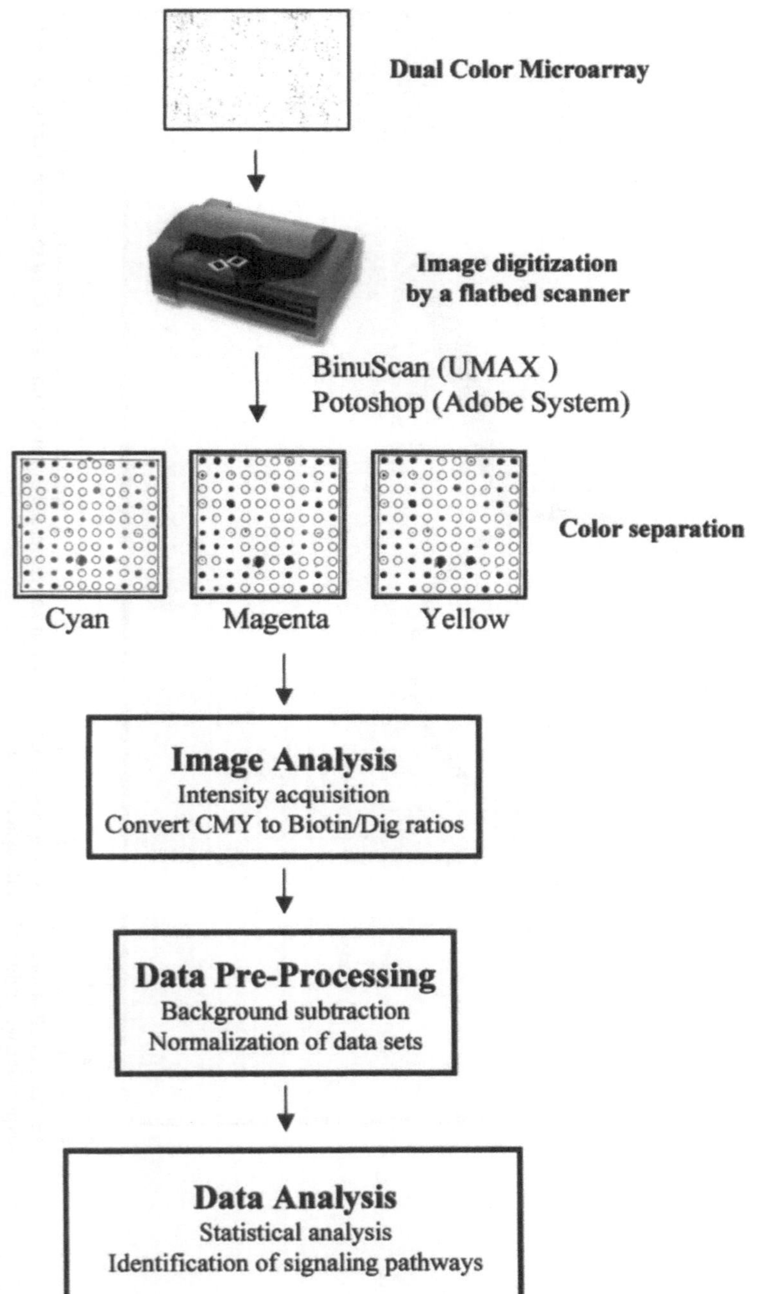

Fig. 3.4. Image and data analysis procedures for cDNA microarray using colorimetric detection method

a threefold expression difference (Chen et al. 1998). Based on our experiments using microarray/CD, cells derived from the same lineage showed that 1–3% of genes had a fivefold expression difference. In comparisons using cancer cell lines derived from different tissues, we isolated 5% or more genes with a fivefold expression difference. Thus, for a 10,000-gene array, 100 differentially expressed genes are identified. The large amount of data generated by the high-density cDNA microarray method places a heavy load on data analysis that still needs to be addressed to unravel the gene regulation mechanisms.

6 Protocol 1: Preparation of cDNA Targets for Microarray Fabrication

6.1 Equipment and Reagents

- Thermowell (skirted thin-wall 96-well plate; Cat. #AB-0800; ABgene, UK)
- 12-channel pipette (CAPPelen Laboratory Technics A/S, Denmark)
- Thermo-Seal (Cat. #AB-0559; ABgene, UK)
- MJ Research PTC-225 Thermal Cycler
- 10× PCR buffer

6.2 Method

1. Thaw the frozen microplates at 37 °C for 10 min.
2. Prepare cocktail of PCR reagents (50 mM Tris-HCl pH 9.1, 16 mM ammonium sulfate, 3.5 mM $MgCl_2$, 150 µg/ml BSA, 0.16 mM dNTP, 0.2 µM primers and enzyme mixture of Klentaq (10 U; Takara, Japan) and DeepVent (1/256 U; New England Biolabs, Beverly, MA). Place 100 µl of the mixture per well in the 96-well microplates.
3. Using a 12-channel pipette, aspirate 2–3 µl of the bacteria stock into the PCR reaction solution and mix well.
4. Seal the plates with the Thermo-Seal film.
5. Perform PCR reactions by MJ Research PTC-225 Thermal Cycler.
6. PCR conditions:
   - 94 °C for 4 min; 58 °C for 1 min; 72 °C for 4 min – for 1 cycle;
   - 94 °C for 30 s; 58 °C for 1 min; 72 °C for 4 min – for 30 cycles;
   - 72 °C for 10 min, then lower the temperature to 4 °C.
7. After the PCR reactions are complete, run agarose gel electrophoresis to check for the PCR amplifications.
8. Add 10 µl of tracking dye (1.36 mg/ml bromophenol blue, 20% glycerol, 1 mM EDTA, 40% DMSO) to each well.

9. Concentrate the PCR solutions to 10 μl volume by heating the solutions to 95 °C by PCR machines.
10. Quickly cool the plates at –20 °C for 10 min.
11. The plates are then ready for arraying.

7 Protocol 2: cDNA Complex Probe Labeling

7.1 Equipment and Reagents

- Random hexamer primer (GIBCO BRL, Cat. #48190–011)
- Reverse transcriptase and 5×buffer (GIBCO BRL, Cat. #18064–014)
- RNase inhibitor (GIBCO BRL, Cat. #10777–019)
- Biotin-16-dUTP (Boehringer Mannheim, Cat. #1093070)
- Dig-11-dUTP (Boehringer Mannheim, Cat. #1558706)
- Control plant mRNAs: HAT22 (*Arabidopsis thaliana* HAT22 homeobox-leucine zipper mRNA), rbcL (ribulose 1,5-bisphosphate carboxylase large subunit), GA4 (*Arabidopsis thaliana* GA4 mRNA), rca (RUBISCO activase precursor), ASA1 (anthranilate synthase alpha 1), ATPS (*Arabidopsis thaliana* ATP sulfurylase), HAT4 (*Arabidopsis thaliana* HAT4 homeobox-leucine zipper mRNA)

7.2 Method

1. Mix 2 μg mRNA, 1 μl of control plant mRNAs for single-color label (HAT22: 1×10^9, rbcL: 5×10^8, GA4: 1×10^8, rca: 5×10^7, ASA1: 1×10^7, ATPS: 5×10^6 molecules/μl), 6 μl of 50 μM random hexamer and DEPC-H_2O to 29 μl final volume. For dual color mode, use 2 μg of mRNA each in biotin or digoxigenin labeling reactions and add control plants mRNA to the two labeling reactions separately by the following formula: (1) biotin labeling: HAT22: 1×10^8, rbcL: 5×10^7, GA4: 2×10^7, rca: 1×10^7, ASA1: 1×10^7, ATPS: 1×10^7, HAT4: 1×10^7/μl. (2) Dig labeling: HAT22: 1×10^7, rbcL: 1×10^7, GA4: 1×10^7, rca: 1×10^7, ASA1: 1×10^8, ATPS: 5×10^7, HAT4: 2×10^7/μl.
2. Heat for 10 min at 70 °C then chill quickly in ice for 5 min.
3. Add 10 μl of 5× first strand buffer, 5 μl of 0.1 M DTT, 1 μl of 25 mM dATP, dCTP, dGTP mixture, 1 μl of 2 mM dTTP, 2 μl of 1 mM biotin-16-dUTP, or Dig-11-dUTP (1 mM), 0.63 μl of 40 U/μl RNAsin and 1.5 μl of Superscript II (reverse transcriptase, GIBCO BRL; 200 U/μl).
4. Mix well and incubate for 10 min at 25 °C, then at 42 °C for 90 min.

5. Stop the reaction by heating to 94 °C for 5 min.
6. Add 5.5 µl of 3 M NaOH and incubate at 50 °C for 30 min.
7. Add 5.5 µl of 3 M CH_3COOH and incubate at 50 °C for 30 min.
8. Precipitate the labeled cDNA by adding 34 µl of water, 50 µl of 7.5 M ammonia acetate, 10 µg of linear polyacrylamide as carrier and 380 µl of absolute alcohol.
9. Incubate the sample for 30 min at –80 °C. Centrifuge at 15,500× *g* for 15 min.
10. Wash the pellet with 1 ml of 70% ethanol and centrifuge at 15,500 *g* for 5 min.
11. Dissolve the pellet in 36 µl of autoclaved H_2O. For dual-color detection, combine the two individually labeled cDNA solutions together.

8 Protocol 3: Array Hybridization and Color Development

8.1 Equipment and Reagents

- EasiSeal (Hybaid, Cat. #HBOSSSEZ1 E)
- Glass slides (Matsunami, S2214, Japan)
- Blocking powder for hybridization (Boehringer Mannheim, Cat. #1096176)
- Bovine serum albumin (Sigma, Cat. #A2153)
- 20× SSC (Amresco, Cat. # 0918S-2–20XPTM5L)
- SDS (Merck, Cat. #113760)
- Salmon sperm DNA (GIBCO BRL)
- Dextran sulfate (Sigma, Cat. #D6001)
- Streptavidin-b-galactosidase (GIBCO BRL, Cat. #19536–010)
- Anti-digoxigenin-AP Fab fragments (Boehringer Mannheim, Cat. #1093274)
- X-gal (GIBCO BRL, Cat. #15520–018)
- Maleic acid (Sigma, Cat. #M1125)
- *N*-lauroylsarcosin (Sigma, Cat. #L5777)
- Fast Red TR/AS-MX substrate kit (PIERCE, Cat. #34034)
- Polyethylene glycol (Sigma, Cat. #P2139)

8.2 Method

1. The filter membrane carrying the 9,600 EST PCR products is prehybridized in 5 ml of 1× hybridization buffer (4× SSC, 0.1% *N*-lauroylsarcosine, 0.02% SDS, 1% BM blocking reagent), and 100 µg/ml salmon sperm DNA at 63 °C for 1.5 h in a Petri dish.

2. Stick one side of an adhesive EasiSeal double-sticky frame to a clean glass slide and place the pre-hybridized array membrane in the center of the frame with the array side facing up.
3. Mix the complex probe with 2 μl of poly-d(A)10 (10 μg/μl) and 2 μl of human Cot-1 DNA (10 μg/μl) (GIBCO BRL) and 40 μl of 2× hybridization buffer to a final volume of 80 μl.
4. Seal the filter membrane and the complex probe in the EasiSeal assembly and denature the complex probe mixture at 95 °C for 5 min and then cool on ice.
5. Incubate at 95 °C for 5 min and then at 63 °C for 12–16 h (overnight).
6. Wash the filter membrane twice with 5 ml of 2× SSC, 0.1% SDS for 5 min at room temperature.
7. Wash three times for 15 min each with 5 ml of 0.1× SSC, 0.1% SDS at 63 °C.
8. Block the filter membrane with 5 ml of 1% BM blocking reagent containing 2% dextran sulfate at room temperature for 1 h.
9. Incubate with 5 ml mixture containing 700× diluted streptavidin-β-galactosidase (1.38 U/ml, enzyme activity) (GIBCO BRL), 10,000× diluted anti-digoxigenin-alkaline phosphatase (0.075 U/ml, enzyme activity) (Boehringer Mannheim), 4% polyethylene glycol 8000 (Sigma), and 0.3% BSA in 1× TBS buffer at room temperature for 2 h. Note: this formula is for dual-color mode. For single-color mode, anti-Dig-AP is not needed and the incubation time can be reduced to 1 h.
10. Wash with 1× TBS buffer three times for 5 min each.
11. Freshly prepare X-gal substrate solution [1.2 mM X-gal, 1 mM $MgCl_2$, 3 mM $K_3Fe(CN)_6$, 3 mM $K_4Fe(CN)_6$ in 1× TBS buffer] by mixing 50 μl of 120 mM X-Gal and 5 ml of X-Gal substrate buffer. Immerse the filter membrane in the X-gal substrate solution for 45 min at 37 °C with gentle shaking.
12. Wash with 1× TBS.
13. Dual-color development: stain the membrane with 5 ml of Fast Red TR/naphthol AS-MX substrate (Pierce, Rockford, IL) at room temperature for 30 min with gentle shaking.
14. Wash with deionized water. Stop the reaction with 1× PBS containing 20 mM EDTA for 20 min.
15. Air-dry the array membrane.

9 Protocol 4: Modified Catalyzed Reporter Deposition (CARD) Method

9.1 Reagents

- Casein (Sigma, Cat. #C8654)
- Tween 20 (Merck, Cat. #822184)

- Streptavidin-peroxidase (BM, Cat. #1089153)
- PEG (Sigma, Cat. #P-2139)
- Streptavidin-β-Gal (GIBCO BRL, Cat. #19536-010)
- X-Gal (GIBCO BRL, Cat. #15520-018)
- Borax (Sigma, Cat. #B-9876)

9.2 Method

1. Hybridize complex probes to the array as described in Protocol 3 up to and including the washing steps.
2. Add blocking buffer (1× PBS, pH 7.4, 0.05% Tween 20, 7% casein) and incubate at room temperature for 1 h.
3. Add streptavidin-HRP (1:1,000) in BSA buffer (PBST + 1% BSA) containing 0.7% casein and 4% PEG. Incubate at room temperature for 1 h.
4. Wash membrane with PBST at room temperature four times for 5 min each.
5. Freshly prepare amplification solution (0.1 M borate buffer, pH 8.5, 0.003% H_2O_2, 15 µl/ml Biotin-tyramide). Note: biotin-tyramide synthesis is based on method B of Bobrow et al. (1989).
6. Aliquot amplification solution in 5-cm Petri dish at 5 ml/dish, then place the array on a nylon membrane in the dish. Incubate for 15 min at room temperature without agitation.
7. Wash the nylon membrane with PBST at room temperature four times for 5 min each.
8. Add streptavidin-β-gal (1:700) in BSA buffer and incubate at room temperature for 1 h.
9. Wash the nylon membrane with PBST at room temperature four times for 5 min each.
10. Develop color by adding substrate: X-Gal substrate buffer 5 ml, 120 mM X-Gal 50 µl. Incubate at 37 °C for 30–45 min.
11. Rinse the nylon membrane in deionized water and then add stopping solution (0.02 M EDTA in PBS).

References

Benson DA, Boguski MS, Lipman DJ, Ostell J (1977) GenBank. Nucleic Acids Res 25:1–6

Bernard K, Auphan N, Granjeaud S, Victorero G, Schmitt-Verhulst AM, Jordan BR, Nguyen C (1996) Multiplex messenger assay: simultaneous, quantitative measurement of expression for many genes in the context of T cell activation. Nucleic Acids Res 24:1435–1442

Bertucci F, Bernard K, Loriod B, Chang YC, Granjeaud S, Birnbaum D, Nguyen C, Peck K, Jordan BR (1999) Sensitivity issues in DNA array-based expression measurements and performance of nylon microarrays for small samples. Hum Mol Genet 8:1715–1722

Bobrow MN, Harris TD, Shaughnessy KJ, Litt GJ (1989) Catalyzed reporter deposition, a novel method of signal amplification. Application to immunoassays. J Immunol Methods 125:279–285
Bobrow MN, Shaughnessy KJ, Litt GJ (1991) Catalyzed reporter deposition, a novel method of signal amplification. II. Application to membrane immunoassays. J Immunol Methods 137:103–112
Boguski MS, Schuler GD (1995) Establishing a human transcript map. Nature Genet 10:369–371
Braun BS, Freiden R, Lessnick SL, May WA, Denny CT (1995) Identification of target genes for the Ewing's sarcoma EWS/FLI fusion protein by representational difference analysis. Mol Cell Biol 15:4623–4630
Cariaso M, Folta P, Lennon G, Wagner M, Kuczmarski T (1999) IMAGEne I: the clustering of ESTs corresponding to known genes. Bioinformatics 15:965–973
Chen, JJ, Wu R, Yang PC, Huang JY, Sher YP, Han MH, Kao WC, Lee PJ, Chiu TF, Chang F, Chu YW, Wu CW, Peck K (1998) Profiling expression patterns and isolating differentially expressed genes by cDNA microarray system with colorimetry detection. Genomics 51(3):313–324
Chu CC, Paul WE (1997) Figl, an interleukin 4-induced mouse B cell gene isolated by cDNA representational difference analysis. Proc Natl Acad Sci USA 94:2507–2512
DeRisi JL, Iyer VR, Brown PO (1997) Exploring the metabolic and genetic control of gene expression on a genomic scale. Science 278:680–686
Eberwine J, Yeh H, Miyashiro K, Cao Y, Nair S, Finnell R, Zettel M, Coleman P (1992) Analysis of gene expression in single live neurons. Proc Natl Acad Sci USA 89:3010–3014
Escribano J, Garcia-Carmona F, Iborra JL, Lozano JA (1984) Kinetic analysis of chemical reactions coupled to an enzymic step. Biochem J 223:633–638
Gossrau R (1978) Azoindoxyl methods for the investigation of hydrolases. IV. Suitability of various diazonium salts (author's translation). Histochemistry 57:323–342
Hellsing K, Richter W (1974) Immunochemical quantitation of dextran by a polymer enhanced nephelometric procedure. J Immunol Methods 5:147–151
Hubank M, Schatz DG (1994) Identifying differences in mRNA expression by representational difference analysis of cDNA. Nucleic Acids Res 22:5640–5648
Iyer VR, Eisen MB, Ross DT, Schuler G, Moore T, Lee JCF, Trent JM, Staudt LM, Hudson J Jr, Boguski MS, Lashkari D, Shalon D, Botstein D, Brown PO (1999) The transcriptional program in the response of human fibroblasts to serum. Science 283:83–87
Kerr MA, Loomes LM, Thorpe SJ (1994) Enzyme-conjugated antibodies. In: Kerr MA, Thorpe R (eds) Immunochemistry LabFax. BIOS Scientific Publishers, Oxford, pp 149–152
Lee N, Zhang SQ, Testa D (1988) A rapid multicolor Western blot. J Immunol Methods 106:27–30
Lennon GG, Auffray C, Polymeropoulos M, Soares MB (1996) The IMAGE consortium: an integrated molecular analysis of genomes and their expression. Genomics 33:151–152
Mathies RA, Peck K, Stryer L (1990) Optimization of high-sensitivity fluorescence detection. Anal Chem 62:1786–1791
Nguyen C, Rocha D, Granjeaud S, Baldit M, Bernard K, Naquet P, Jordan BR (1995) Differential gene expression in the murine thymus assayed by quantitative hybridization of arrayed cDNA clones. Genomics 29:207–216
Salunga RC, Guo H, Luo L, Bittner A, Joy KC, Chambers JR, Wan JS, Jackson MR, Erlander MG (1999) Gene expression analysis via cDNA microarrays of laser capture microdissected cells from fixed tissue. In: Schena M (ed) DNA microarrays. A practical approach. Oxford University Press, Oxford, pp 121–137
Schena M, Shalon D, Davis RW, Brown PO (1995) Quantitative monitoring of gene expression patterns with a complementary DNA microarray. Science 270:467–470
Schuler GD, Boguski MS, Stewart EA, Stein LD, Gyapay G, Rice K, White RE, Rodrigues-Tome P, Aggarwal A, Barjorek E et al. (1996) A gene map of the human genome. Science 274:540–546
Thorpe SJ, Kerr MA (1994) Common immunological techniques. In: Kerr MA, Thorpe R (eds) Immunochemistry LabFax. BIOS Scientific Publishers, Oxford, pp 198–204
Tijssen P (1985) Practice and theory of enzyme immunoassays, Elsevier, Amsterdam
Van Gelder RN, von Zastrow ME, Yool A, Dement WC, Barchas JD, Eberwine JH (1990) Amplified RNA synthesized from limited quantities of heterogeneous cDNA. Proc Natl Acad Sci USA 87:1663–1667
Zaitsu K, Ohkura Y (1980) New fluorogenic substrates for horseradish peroxidase: rapid and sensitive assays for hydrogen peroxide and the peroxidase. Anal Biochem 109:109–113
Zhang L, Zhou W, Velculescu VE, Kern SE, Hruban RH, Hamilton SR, Vogelstein B, Kinzler KW (1997) Gene expression profiles in normal and cancer cells. Science 276:1268–1272

CHAPTER 4

cDNA Macroarrays and Microarrays on Nylon Membranes with Radioactive Detection

BÉATRICE LORIOD[1], GENEVIÈVE VICTORERO[1]
and CATHERINE NGUYEN[1]

[1]TAGC/CIML, Parc Scientifique de Luminy, Case 906, 13288 Marseille, Cedex 9, France

1 Introduction

As more and more genes are identified, efforts are redirected towards understanding the precise temporal and cellular control of gene expression. Expression measurements are a key step in this direction, and can be performed on a reasonably large scale using highly parallel hybridisation methods (Jordan 1998; Granjeaud et al. 1999; Supplement to *Nature Genetics* 1999). The method we describe here differs from others in two respects: the nature of the solid support (nylon) and the label used for complex probes (radioactive isotopes). Large-scale hybridisation techniques on nylon membranes to measure gene expression are presently available in two flavours: bacterial colonies or PCR products on high-density membranes, on the one hand, and microarrays of PCR products on the other. Microarrays are essentially a miniaturisation of the long-standing high-density membranes where overall dimensions and interelement spacing have typically been brought down five- to ten-fold. In both cases expression measurements are obtained after hybridisation with radioactive complex probes.

One of the crucial issues in array technology is the ability to increase signal-to-noise ratio in order to drive down the detection threshold and reliably assay genes with low expression. There are five parameters to take into account: the amount of target (cDNA bound to the support), the complex probe concentration, the specific activity of the labelled material, the duration of hybridisation and exposure. Improving the last three parameters is relatively straightforward, and there is no theoretical limitation to the probe concentration: this is in practice only limited by the amount of sample available (biopsies, sorted cells, etc.). However, the intrinsic physical characteristics of nylon membranes allow a significant increase in the amount of immobilised target, hence improving overall sensitivity. In fact, when working with a limited amount of RNA, nylon microarrays coupled with radioactive labelling seem to be the method of choice (Bertucci et al. 1999).

There are three essential elements necessary to successfully implement this technology: the robot (for spotting colonies or PCR products), the imaging plate device (for image capture, often referred to as a phosphor imager), and the image analysis software (signal quantification).

2 Manufacture of Nylon Arrays

2.1 Clone Resources

Most of the cDNA clones used here to make nylon arrays come from cDNA libraries of the IMAGE consortium. The IMAGE clone bank and the associated EST database have become essential tools for the research community, and resource centres in Europe such as the German Human Genome Project, RZPD (http://www.dkfz-heidelberg.de), and the UK HGMP Resource Centre Hinxton (http://www.hgmp.mrc.ac.uk) are committed to maintaining and extending their utility. Working with IMAGE clones requires caution, because a proportion of the clones have recently been shown to be contaminated with a bacteriophage, apparently coliphage T1-related in most if not all cases. Phage T1 contamination can spread rapidly to other *E. coli* cultures and ruin whole libraries. However, most resource centres have put procedures in place to minimise the chance of transfer of phage contamination to users through IMAGE shipments. In addition, we use our own cDNA libraries (http://tagc.univ-mrs.fr/resources/library.html), and for some additional clones [including the control *Arabidopsis thaliana* gene: chlorophyll synthetase (*CG03*)] the cDNA insert was transferred from the original cloning vector to that used for the cDNA library (the same plasmid vector and bacteria). Three clones containing essentially polyA sequence (50, 60, 90 bp) were obtained by appropriate digests of the polyA tail of sequenced cDNAs, followed by cloning at the multiple cloning site of the pT7T3D-Pac vector. The exogenous cDNA from *Arabidopsis thaliana*, polyA and pure vector clones represent the minimum set of controls needed to properly analyse and compare independent measurements.

2.2 Arrays on Nylon Membranes

Macroarrays are nylon (or other) membranes that carry bacterial colonies or PCR products with a typical diameter of 0.8–1.5 mm and a pitch (centre-to-centre spacing) of approximately 2 mm. Maximum densities obtainable are

---

Fig. 4.1. Common primers we designed for the three vectors commonly used in IMAGE libraries: **A** Lafmid, **B** Bluescript and **C** pT7T3. Sequence of multiple cloning site and 100 bp around it are presented. *Boxes* define positions of different primers and corresponding sequences are shown in *bold*. The first PCR is done with the most exterior primers. LBP 2S and LBP 2AS primers are designed to be used if nested PCR is needed. LBP 3 is used as vector probe to evaluate the amount of cDNA fixed onto filters. The set Oligo 15 and Oligo 16 can also be used but not T3, T7 and T7 bis only found on Bluescript and Pac-pT7T3

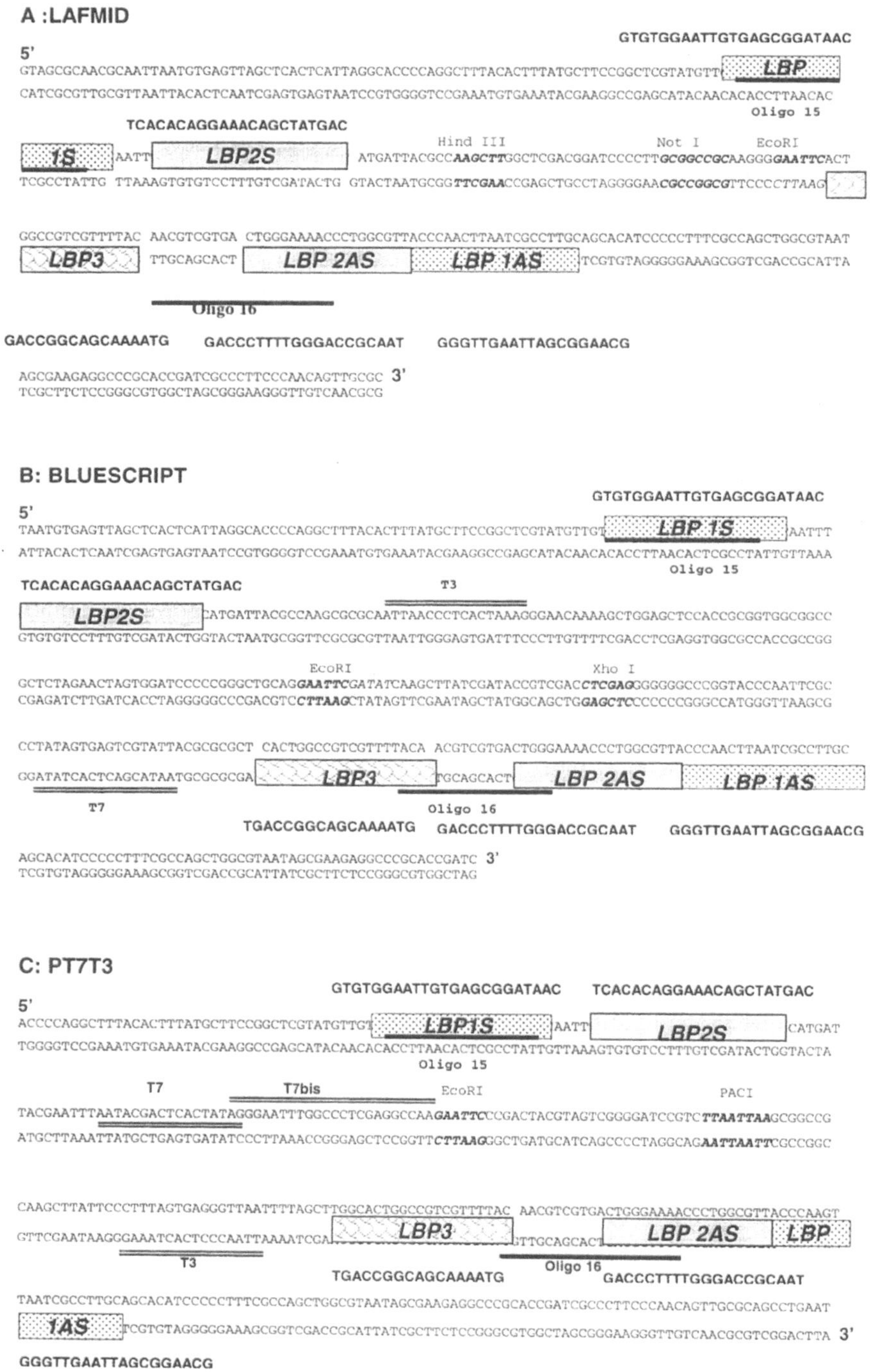

A :LAFMID

GTGTGGAATTGTGAGCGGATAAC

5'
GTAGCGCAACGCAATTAATGTGAGTTAGCTCACTCATTAGGCACCCCAGGCTTTACACTTTATGCTTCCGGCTCGTATGTT LBP
CATCGCGTTGCGTTAATTACACTCAATCGAGTGAGTAATCCGTGGGGTCCGAAATGTGAAATACGAAGGCCGAGCATACAACACACCTTAACAC
Oligo 15

TCACACAGGAAACAGCTATGAC

Hind III Not I EcoRI

1S AATT LBP2S ATGATTACGCC**AAGCTT**GGCTCGACGGATCCCCTT**GCGGCCGC**AAGGG**GAATTC**ACT
TCGCCTATTG TTAAAGTGTGTCCTTTGTCGATACTG GTACTAATGCGG**TTCGAA**CCGAGCTGCCTAGGGGAA**CGCCGGCG**TTCCC**CTTAAG**

GGCCGTCGTTTTAC AACGTCGTGA CTGGGAAAACCCTGGCGTTACCCAACTTAATCGCCTTGCAGCACATCCCCCTTTCGCCAGCTGGCGTAAT
LBP3 TTGCAGCACT LBP 2AS LBP 1AS TCGTGTAGGGGGAAAGCGGTCGACCGCATTA

Oligo 16

GACCGGCAGCAAAATG GACCCTTTTGGGACCGCAAT GGGTTGAATTAGCGGAACG

AGCGAAGAGGCCCGCACCGATCGCCCTTCCCAACAGTTGCGC 3'
TCGCTTCTCCGGGCGTGGCTAGCGGGAAGGGTTGTCAACGCG

B: BLUESCRIPT

GTGTGGAATTGTGAGCGGATAAC

5'
TAATGTGAGTTAGCTCACTCATTAGGCACCCCAGGCTTTACACTTTATGCTTCCGGCTCGTATGTTGT LBP 1S AATTT
ATTACACTCAATCGAGTGAGTAATCCGTGGGGTCCGAAATGTGAAATACGAAGGCCGAGCATACAACACACCTTAACACTCGCCTATTGTTAAA
Oligo 15

TCACACAGGAAACAGCTATGAC T3

LBP2S CATGATTACGCCAAGCGCGCAATTAACCCTCACTAAAGGGAACAAAAGCTGGAGCTCCACCGCGGTGGCGGCC
GTGTGTCCTTTGTCGATACTGGTACTAATGCGGTTCGCGCGTTAATTGGGAGTGATTTCCCTTGTTTTCGACCTCGAGGTGGCGCCACCGCCGG

EcoRI Xho I
GCTCTAGAACTAGTGGATCCCCCGGGCTGCAG**GAATTC**GATATCAAGCTTATCGATACCGTCGAC**CTCGAG**GGGGGGCCCGGTACCCAATTCGC
CGAGATCTTGATCACCTAGGGGGCCCGACGTC**CTTAAG**CTATAGTTCGAATAGCTATGGCAGCTG**GAGCTC**CCCCCCGGGCCATGGGTTAAGCG

CCTATAGTGAGTCGTATTACGCGCGCT CACTGGCCGTCGTTTTACA ACGTCGTGACTGGGAAAACCCTGGCGTTACCCAACTTAATCGCCTTGC
GGATATCACTCAGCATAATGCGCGCGA LBP3 TGCAGCACT LBP 2AS LBP 1AS
T7 Oligo 16
TGACCGGCAGCAAAATG GACCCTTTTGGGACCGCAAT GGGTTGAATTAGCGGAACG

AGCACATCCCCCTTTCGCCAGCTGGCGTAATAGCGAAGAGGCCCGCACCGATC 3'
TCGTGTAGGGGGAAAGCGGTCGACCGCATTATCGCTTCTCCGGGCGTGGCTAG

C: PT7T3

GTGTGGAATTGTGAGCGGATAAC TCACACAGGAAACAGCTATGAC

5'
ACCCCAGGCTTTACACTTTATGCTTCCGGCTCGTATGTTGT LBP1S AATT LBP2S CATGAT
TGGGGTCCGAAATGTGAAATACGAAGGCCGAGCATACAACACACCTTAACACTCGCCTATTGTTAAAGTGTGTCCTTTGTCGATACTGGTACTA
Oligo 15

T7 T7bis EcoRI PACI
TACGAATTTAATACGACTCACTATAGGGAATTTGGCCCTCGAGGCCAA**GAATTC**CCGACTACGTAGTCGGGGATCCGTC**TTAATTAA**GCGGCCG
ATGCTTAAATTATGCTGAGTGATATCCCTTAAACCGGGAGCTCCGGTT**CTTAAG**GGCTGATGCATCAGCCCCTAGGCAG**AATTAATT**CGCCGGC

CAAGCTTATTCCCTTTAGTGAGGGTTAATTTTAGCTTGGCACTGGCCGTCGTTTTAC AACGTCGTGACTGGGAAAACCCTGGCGTTACCCAAGT
GTTCGAATAAGGGAAATCACTCCCAATTAAAATCGA LBP3 TTGCAGCACT LBP 2AS LBP
T3 Oligo 16
TGACCGGCAGCAAAATG GACCCTTTTGGGACCGCAAT

TAATCGCCTTGCAGCACATCCCCCTTTCGCCAGCTGGCGTAATAGCGAAGAGGCCCGCACCGATCGCCCTTCCCAACAGTTGCGCAGCCTGAAT
1AS TCGTGTAGGGGGAAAGCGGTCGACCGCATTATCGCTTCTCCGGGCGTGGCTAGCGGGAAGGGTTGTCAACGCGTCGGACTTA 3'
GGGTTGAATTAGCGGAACG

thus of the order of 36 spots/cm^2; typical macroarrays may measure 8×12 cm^2 (microtitre plate size) and carry a few thousand spots. They are normally used with radioactive complex probes in large hybridisation volumes (several millilitres), quantitative results being acquired with a phosphor screen system. Two difficulties can affect the quality of colony macroarrays in particular regarding data interpretation: the bacteriophage contamination already discussed above and the possible cross contamination between clones which can take place during plate manipulation. PCR macroarrays eliminate part of the problem since PCR products can be stored and re-amplified, as well as checked by gel migration to identify cross-contamination before array manufacture. One of the major problems with PCR production is to obtain sufficient amounts of amplified DNA and to avoid large clone-to-clone variations: this is far from obvious. We occasionally observed large variations in the amounts of PCR product between clones, related to insert size and sequence as well as sometimes to trivial factors such as the position of tubes on the PCR machines (primers are described in Fig. 4.1).

Microarrays (which carry only PCR products) correspond to much smaller dimensions, with spot diameters of ~200 μm and a pitch of 300–500 μm. Maximum densities can reach several thousand spots per square centimetre; typical microarrays carry up to 10,000 spots in a surface corresponding to a microscope slide (7.3×1.8 cm). Figure 4.2 shows two examples of microarrays containing cDNAs on a membrane hybridised with a vector oligonucleotide and a complex probe. The vector hybridisation (which reveals the amount of DNA on the membrane) clearly shows the difference in quality between the two microarrays, only the one with even deposits being acceptable for quantitative experiments.

Arrays on nylon membranes present some advantages. The blotting substrate is well known and controlled, and covalent binding of DNA to positively charged nylon membranes is strong enough to withstand stringent stripping so that arrays can be re-probed several times without significant degradation of signal (at least five times). More importantly, a large amount of target DNA can be fixed on to nylon, due to the large binding available surface per square millimetre. Nylon membranes are open-pore sponge-like structures with finite thickness (~100 μm); Figure 4.3 shows the membrane surface structure analysed by atomic force microscopy (AFM) (kindly provided by M. Marylley and J. Thimonier, Université de la Méditerranée, Marseille). According to the manufacturer of Hybond N (Amersham Pharmacia Biotech), the binding capacity of these membranes is 600 μg/cm^2, which means that approximately 200 ng of DNA could be bound per spot of 200 μm diameter (31.4×10^3 μm^2). Another decisive advantage of nylon is that purification of the PCR products is not needed, contrary to glass slide supported arrays.

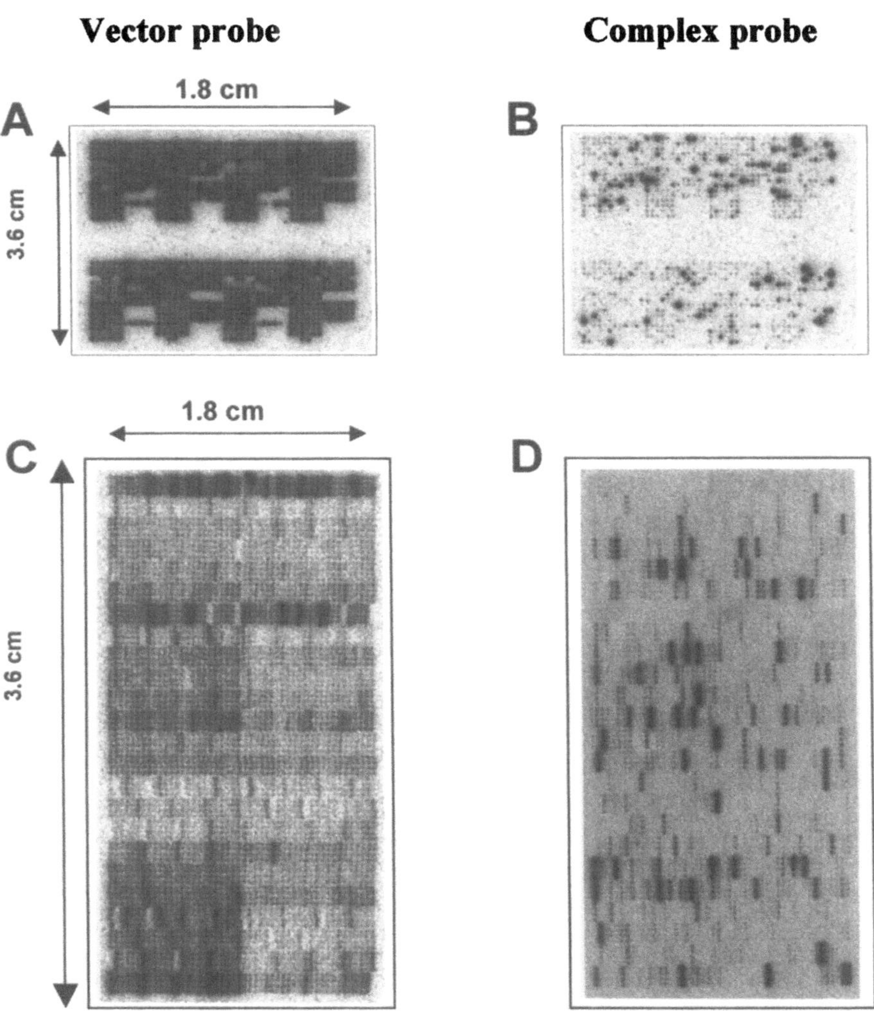

Fig. 4.2. Nylon microarrays hybridised with radiolabelled probes. Microarrays using PCR products from twelve 96-well plates were spotted singly (**A**, **B**) and in quadruplicate with GMS 417 arrayer, yielding respectively 1,152 and 4,608 spots (375-μm pitch) on membranes. **A** and **C** Hybridisation with vector oligonucleotide to determine amount of DNA in each spot. Comparison between these two images shows that DNA amounts are very homogeneous in **A** but quite variable in **C**, reflecting the care taken in PCR product preparation and standardisation. Arrays such as the one shown in **C** give unreliable expression data for genes corresponding to spots containing small amounts of DNA. **B** and **D** Hybridisation with complex probes shows great variation of expression patterns and brings out typical quadruplicate local patterns on filter **D**. Note that complex probe hybridisation does not allow the evaluation of the quantities of the cDNA deposited on filters which can lead to artefactual spot-to-spot variation of measured expression levels if data are not corrected

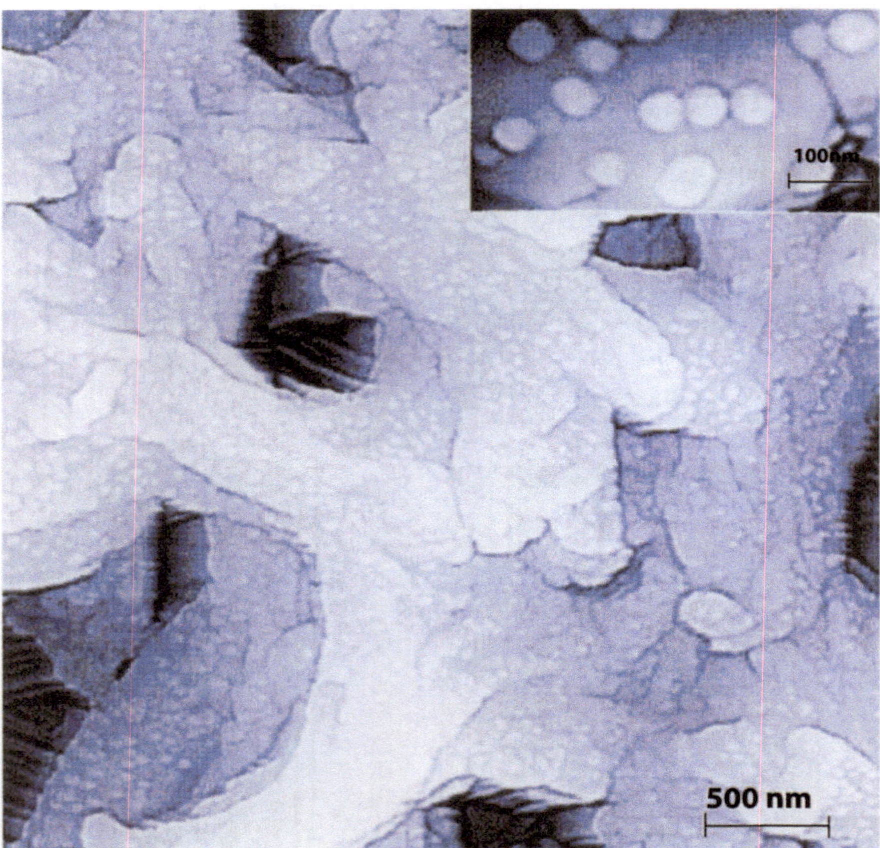

Fig. 4.3. Membrane structure. Picture has been obtained by atomic force microscopy (AFM) with gold colloidal beads of 20-nm diameter. Beads appear as small protrusions stuck to membrane fibers. Membrane looks like a three-dimensional labyrinth. *Upper right corner* Close-up of membrane fiber covered with 20-nm beads (to be compared to DNA thickness of 2 nm). Direct observation of DNA strands has proven impossible using AFM due to irregularities of nylon fiber surface. (Experiments performed and image provided by M. Marilley and J. Thimonier, Université de la Méditerranée, Marseille)

2.3 The Arraying Machine

We currently use two solid pin arrayers: the Biogrid from Bio Robotics (http://www.biorobotics.com) and the GMS 417 microarraying system (http://www.affymetrix.com). Our version of the Biogrid is used for macroarrays; the tools carry either 96 or 384 pins, which compensates for the relatively slow spotting rate resulting from the numerous back-and-forth movements between reagent reservoir and the membranes. The GMS 417 is used for microarrays, and uses a "pin and ring" system for deposition. Small rings dip into microtiter plate wells and take up a meniscus of DNA solution that contains around 1 µl; a vertical pin moves through the meniscus to deposit liquid on the array. A pin punch deposits ~0.5 nl of product per hit. This allows a theoretical maximum of 2,000 hits per ring charge. Actually, ~400 hits can be practically achieved (20% of the PCR product is really used) before the drop pops out of the ring, mostly due to slow evaporation during the long run, this in spite of hygrometry control. The pin and ring concept confers to this arrayer a high speed rate, this advantage being moderated by the fact that our equipment version only carries four pins and rings and by the waste of the PCR product per run. Whatever system is used, the amount of reagent deposited depends on the pin diameter (see Table 4.1).

Table 4.1. Arrayer parameters. Pin diameters (according to manufacturers) of arrayers used are shown in column 2. Volume taken up by pin (column 3) and speed (column 7) are estimated according to our own experience with the robot. Average spot size (apparent) and interspot spacing are calculated after vector hybridisation and image capture with the help of Fuji software. Deposited amounts (column 5) are calculated from volume and sample concentration, or measured for colonies (see Nguyen et al.. 1995)

Macroarrays with BIOGRID arrayer	**Pin diameter**	**Volume**	**Average spot size** on nylon	**Amount expected (sample conc. 1 µg/µl)**	**Interspot spacing**	**Approximate speed (spots/min)**
96-pin head on 24 membranes	0.7 mm	3 nl	1.4 mm	3 ng	2 mm	250
384-pin head on 24 membranes	0.4 mm	1.5 nl	1.2 mm	1.5 ng	1—2 mm	1,000
Bacteria			1.2 mm	100 to 200 ng plasmid/ colony, i.e. 20—60 ng of insert		
Microarrays with GMS417 arrayer						
Pin and ring system 4-pin head on 42 slides	125 µm	0.6 nl	300 µm	0.6 ng	375 µm	960

3
Labelling and Hybridisation with Complex Probes from Total RNA

Reverse transcriptase (RT), primed by oligo-dT, simultaneously synthesises and labels single-stranded DNA from total RNA. To compare independent hybridisations with precision, a known amount of *CG03* mRNA, in vitro transcribed with T3 polymerase from the corresponding cDNA clone (from *Arabidopsis thaliana* library), is added before labelling to the total RNA of each cell type or tissue to analyse (Bernard et al. 1996). Complex probes are prepared from total RNA plus *CG03* mRNA with an excess of oligo-dT(25) in order to saturate the polyA tails and ensure that reverse transcription will start near the beginning of the polyA tail, and thus avoid products containing long poly T stretches. Before hybridisation any remaining poly T regions are blocked with an excess of oligo-dA(80).

Originally, hybridisation was performed with dCTP-a^{32}P; however, we observed that weak signals from some spots could be adversely affected by spillover from strong signals of neighbouring spots because of the high energy emission of ^{32}P (Beta Max energy: 1.709). The separation between neighbouring spots was clearly improved with the use of ^{33}P which has a lower energy (Beta Max energy: 0.248), as shown in Fig. 4.4.

The quality of the RNA sample can have significant effect on the final signals observed and hence the degradation state should be evaluated. Classically, this is evaluated on gel: a rule of thumb is that the resulting signal intensity of the 28 S band must be twice that of the 18 S band. Often this ratio is not exactly observed and calls for further evaluation. Different states of degradation can be seen and evaluated on Northern blots of total RNA after hybridisation with an oligonucleotide specific for 28 S RNA, a sensitive indicator for degradation if the analysis is performed under denaturing conditions. Degradation is evaluated by the appearance and intensity of lower molecular weight bands. Figure 4.5 shows the consequence to expression profiling results after hybridisation performed with complex probes made from an RNA sample which was gradually degraded. This shows that although slight degradation is acceptable, extensively degraded RNA is unusable. Because this evaluation costs RNA, it may be impossible to apply this test in some instances where samples are precious and limited (biopsies, sorted cells). New technologies such as "lab-on-chip" microcapillary electrophoresis (Agilent, http://www.chem.agilent.com) can yield similar data with far less starting material.

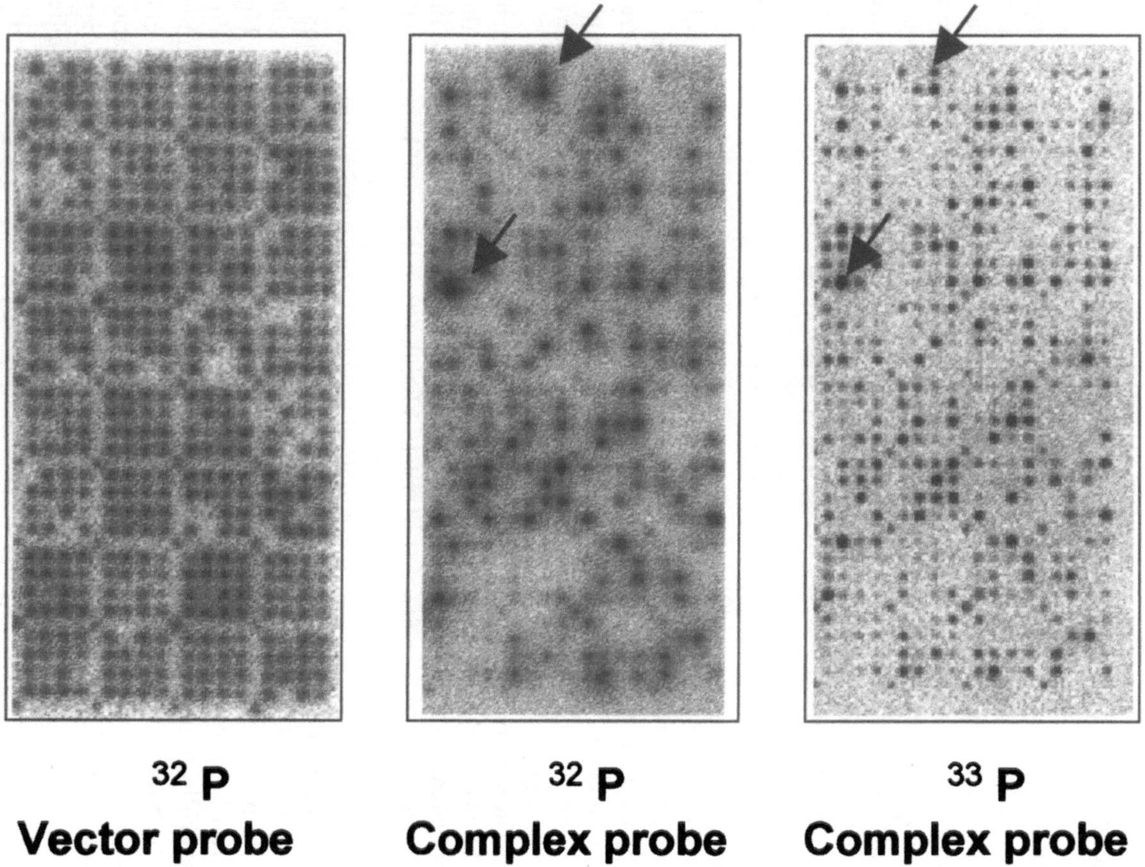

Fig. 4.4. Effect of radioisotope emission energy. Colony macroarrays of 8×12 cm were hybridised with vector oligonucleotide (*left*), with complex probes made from 25 µg of total RNA labelled with a-^{32}PdCTP (*centre*) and with a-^{33}PdCTP (*right*). Fairly homogeneous signals are observed after hybridisation with vector probes under saturation conditions (see Protocols), unlike signals from complex probes that are correlated with gene expression levels. Highly expressed transcripts provide strong signals (*arrows*) which clearly pollute neighbouring spots in the case of ^{32}P (*centre*). This artefact is essentially eliminated by use of ^{33}P

4 Signal Analysis

4.1 Detection and Quantification of Hybridisation Results

Quantitative data are obtained using an imaging plate device. The hybridised membrane is exposed to an imaging plate for 24–72 h and then scanned in a Fuji BAS 1500 (100-μm pixel size) or Fuji BAS 5000 (25-μm pixel size) system (http://www.fujifilm-europe.com/i). This system is far superior in quantitative applications to simpler methods based on exposure to X-ray film, because of their lack of linearity. Nonetheless, scanning of X-ray films with subsequent processing has been used with some success (Gress et al. 1992) in spite of its limitations. At this time, imaging plate systems are most popular. Their resolution (25–200 μm pixel size, depending on make and model) is adequate, and their range of linear response extends over 4–5 decades (Johnston et al. 1990; Moryand Hamaoka 1994). The standard software provided with these machines, however, is not adapted to quantitative analysis of high-density membranes. In our group, hybridisation signals are quantified by the Bio Image software running on a UNIX workstation (Genomic Solutions, USA; http://www.genomicsolutions.com; Patterson and Latter 1993). The resulting quantified data are then analysed on a microcomputer using EXCEL software with macro-commands that compute values for each cDNA (see Fig. 4.6). This procedure is described in Granjeaud et al. 1996. The successive operations are briefly described in Section 9 of this chapter. In addition, a number of software tools have been developed or are under development, incorporating for example the hierarchical clustering methods described in Eisen et al. (1998). A second essential task is to keep track of information acquired on hundreds or thousands of clones in an effective and user-friendly fashion. To this effect, we developed a laboratory notebook system called "LABNOTE" (Imbert et al. 1999). This laboratory database has been constructed in the 4th Dimension (ACI) relational database management system and is presently available to academic users (http://tagc.univ-mrs.fr/pub/labnote).

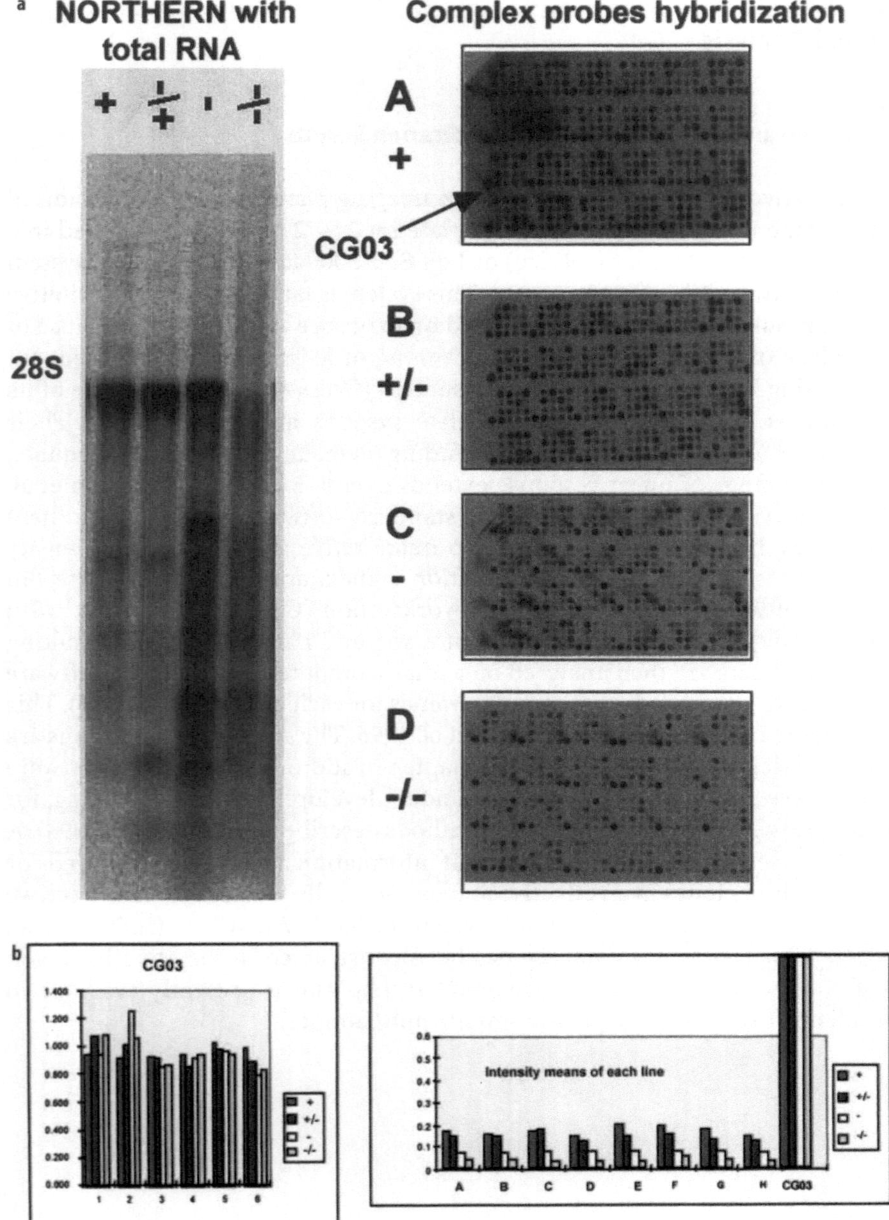
a
NORTHERN with total RNA
Complex probes hybridization
+
+/-
-
-/-
28S
A
+
CG03
B
+/-
C
-
D
-/-
b
CG03
Intensity means of each line
A
B
C
D
E
F
G
H
CG03

4.2 Sensitivity

Most of the applications of cDNA arrays require a large amount of RNA and are thus limited to experiment with material derived from abundant sources. We have shown that actual sensitivity, defined as the amount of sample necessary for detection of a given mRNA species, is similar for microarrays on glass slides and oligonucleotide chips with fluorescence detection, as well as for nylon microarrays with colorimetric detection (Chen et al. 1998). For the two first systems this is mainly due to the very different amounts of target present on the respective arrays. For the last one, the limited sensitivity of colorimetric detection offsets the advantage of high target amounts and small hybridisation volumes, leading to a figure similar to the two other methods. The combination of nylon microarrays with ^{33}P-labelled radioactive probes provides 100-fold better sensitivity (Bertucci et al. 1999), making it possible to perform expression profiling experiments using submicrogram amounts of unamplified total RNA from small biological samples.

4.3 Measurement Parameters, Treatment of Artefacts, Normalisation

Assuming that probe sequences represent the labelled RNA population and that targets represent cDNA spotted on membranes, for probe sequences present in small amounts relative to their targets, and if the hybridisation reaction is far from completion as described by Beltz et al. (1983), the signal is proportional to both target amount and abundance of a given probe species. In our membranes, amounts of plasmid DNA vary according to filter type: the amount on colony membranes are 100–200 ng per colony, i.e., 20–60 ng of insert material, and for PCR membranes the amounts are from 3–0.6 ng for

Fig. 4.5. Effect of RNA degradation. **a** (*left*) Northern blot done with same RNA sample subjected to progressive degradation from intact (+) to completely degraded (–/–) was hybridised with labelled 28S oligonucleotide. Resulting image shows 28S band which becomes progressively weaker and finally disappears, while additional faster-migrating bands become stronger before complete degradation. *Right* Image detail from four high-density filters hybridised with complex probes made with these RNAs spiked with same amount (2 ng) of same batch of exogenous *CG03* RNA (from *Arabidopsis thaliana*). *CG03* signals (*bottom left corner* of each square pattern of 16 spots) can be easily individualised on left edge of these filters and appear fairly homogeneous between the four hybridisations. **b** Spots were quantified and results are shown. *Left* Histogram representing *CG03* mean results from two columns of each hybridisation. Signals are homogeneous across each filter and between filters. *Right* Same representation of mean intensities for each line of cDNA clones. A clear gradual global signal decrease is observed; nonetheless, we observed that individual signals do not change in a homogeneous manner, thus leading to false ratios that can lead to erroneous data interpretation

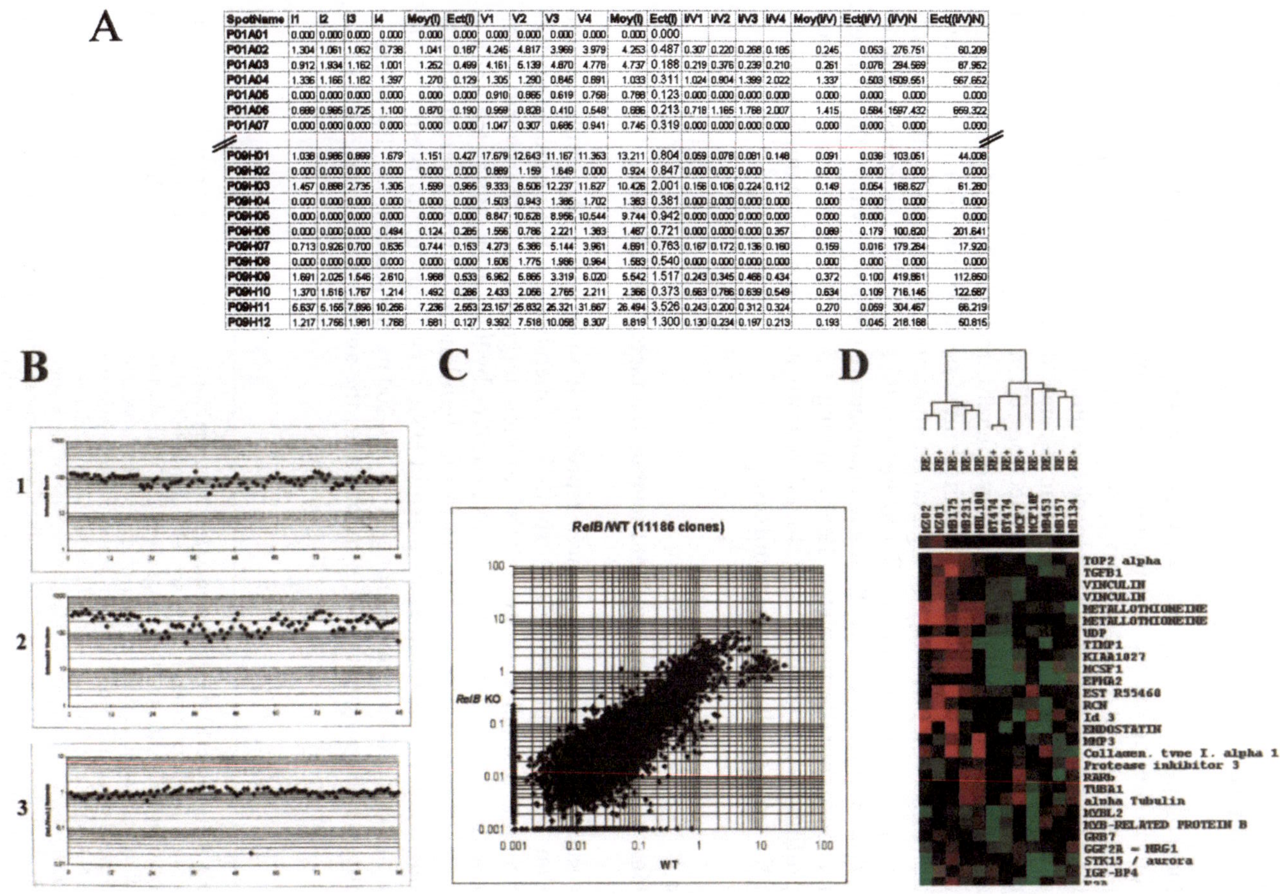

SpotName	I1	I2	I3	I4	Moy(I)	Ect(I)	V1	V2	V3	V4	Moy(I)	Ect(I)	I/V1	I/V2	I/V3	I/V4	Moy(I/V)	Ect(I/V)	(I/V)N	Ect((I/V)N)
P01A01	0.000	0.000	0.000	0.000	0.000	0.000	0.000	0.000	0.000	0.000	0.000	0.000								
P01A02	1.304	1.061	1.062	0.738	1.041	0.187	4.245	4.817	3.969	3.979	4.253	0.487	0.307	0.220	0.268	0.186	0.245	0.053	276.751	60.209
P01A03	0.912	1.934	1.162	1.001	1.252	0.499	4.161	5.139	4.870	4.778	4.737	0.188	0.219	0.376	0.239	0.210	0.261	0.078	294.569	87.952
P01A04	1.336	1.166	1.182	1.397	1.270	0.129	1.305	1.290	0.845	0.891	1.033	0.311	1.024	0.904	1.399	2.022	1.337	0.503	1509.551	567.652
P01A05	0.000	0.000	0.000	0.000	0.000	0.000	0.910	0.865	0.619	0.768	0.788	0.123	0.000	0.000	0.000	0.000	0.000	0.000	0.000	0.000
P01A06	0.689	0.965	0.725	1.100	0.870	0.190	0.959	0.828	0.410	0.548	0.686	0.213	0.718	1.165	1.768	2.007	1.415	0.584	1597.432	659.322
P01A07	0.000	0.000	0.000	0.000	0.000	0.000	1.047	0.307	0.685	0.941	0.745	0.319	0.000	0.000	0.000	0.000	0.000	0.000	0.000	0.000
//																				//
P09H01	1.038	0.986	0.899	1.679	1.151	0.427	17.679	12.643	11.167	11.363	13.211	0.804	0.069	0.078	0.081	0.148	0.091	0.039	103.061	44.008
P09H02	0.000	0.000	0.000	0.000	0.000	0.000	0.889	1.159	1.649	0.000	0.924	0.847	0.000	0.000	0.000		0.000	0.000	0.000	0.000
P09H03	1.457	0.898	2.735	1.306	1.599	0.966	9.333	8.506	12.237	11.627	10.426	2.001	0.156	0.106	0.224	0.112	0.149	0.054	168.627	61.280
P09H04	0.000	0.000	0.000	0.000	0.000	0.000	1.503	0.943	1.386	1.702	1.383	0.381	0.000	0.000	0.000	0.000	0.000	0.000	0.000	0.000
P09H05	0.000	0.000	0.000	0.000	0.000	0.000	8.847	10.628	8.956	10.544	9.744	0.942	0.000	0.000	0.000	0.000	0.000	0.000	0.000	0.000
P09H06	0.000	0.000	0.000	0.494	0.124	0.285	1.566	0.786	2.221	1.383	1.487	0.721	0.000	0.000	0.000	0.357	0.089	0.179	100.820	201.641
P09H07	0.713	0.926	0.700	0.635	0.744	0.153	4.273	6.386	5.144	3.961	4.891	0.763	0.167	0.172	0.136	0.160	0.159	0.016	179.284	17.920
P09H08	0.000	0.000	0.000	0.000	0.000	0.000	1.606	1.775	1.986	0.964	1.583	0.540	0.000	0.000	0.000	0.000	0.000	0.000	0.000	0.000
P09H09	1.691	2.025	1.646	2.610	1.968	0.533	6.962	5.866	3.319	6.020	5.542	1.517	0.243	0.345	0.468	0.434	0.372	0.100	419.861	112.860
P09H10	1.370	1.616	1.767	1.214	1.492	0.286	2.433	2.056	2.765	2.211	2.366	0.373	0.563	0.786	0.639	0.549	0.634	0.109	716.145	122.687
P09H11	6.637	6.155	7.896	10.256	7.236	2.553	23.157	25.832	25.321	31.667	26.494	3.526	0.243	0.200	0.312	0.324	0.270	0.059	304.467	66.219
P09H12	1.217	1.756	1.981	1.768	1.681	0.127	9.392	7.518	10.058	8.307	8.819	1.300	0.130	0.234	0.197	0.213	0.193	0.045	218.188	50.815

macro- and microarrays respectively depending on the spotting system used (see Table 4.1). We have shown in previous work (Nguyen et al. 1995; Bertucci et al. 1999) that a maximum of 20% of the probe is bound to the immobilised targets; this is true for abundant species (1%) as well as for the rare transcripts (<0.01%) ensuring target excess. This is particularly important because of the frequent need to reduce the amount of RNA used, yielding proportionally reduced intensities, and thus burying weaker signals under the detection threshold.

We have shown that the hybridisation signal increases linearly with the amount of spotted target (Nguyen et al. 1995; Bertucci et al. 1999). This result, which was not a priori obvious under these conditions, allows signals to reach higher values and consequently permits minimisation of the amounts of RNA used for analysis. This point is particularly important when working with precious biological micro samples. In addition, because signal increases linearly with the target amount, hybridisation signals should be corrected for the differential amount of cDNA spotted. This can be achieved using data obtained with a vector probe. Correcting hybridisation signals by target amount information from vector hybridisation permits accurate differential analysis between distinct cDNAs present on the same membrane hybridised with a single complex probe (intergene comparisons). This is particularly important in differential screening when clones are selected on the basis of their expression inside a chosen interval level. Similarly, when comparing between conditions, this correction (among others, see below about spike RNA) allows accurate differential measurement across multiple independent hybridisations since values obtained are "absolute" (to a certain extent).

Of course, hybridisation signal increases linearly with the complex probe amount, and thus signal intensity is proportional to the abundance of the corresponding sequences in the sample. As mentioned earlier, the five main parameters influencing signal are: amount of cDNA immobilised on support, amount of labelled RNA, specific activity, hybridisation kinetics and exposure

◄

Fig. 4.6. Data treatment. A Data analysis (report) from UNIX station after transfer to Excel and formatting with macro-commands: table presented here is ready for further analysis. Spot names correspond to cDNA position in plates; *I1—I4* and *V1—V4* are intensity measures with complex probes and vector probes respectively (each in quadruplicate on filters); *Moy(I)* and *Ect(I)* are means and standard deviations respectively. *I/V1—I/V4* are intensity data corrected by vector hybridisation, then *Moy(I/V)* and *Ect(I/V)* are means and standard deviations. Last two columns correspond to data normalised by the exogenous spike and its associated standard deviation. **B** *1 CG03* signals with spiked complex probes; *2* with vector probes; *3* values obtained after vector correction showing that part of observed dispersion with complex probe is due to different amounts of cDNA spotted on membranes. **C** One example of a quantitative differential screen: after normalisation, data from two independent hybridisations are compared in order to select genes differentially expressed. Usually, clones displaying more that two standard deviations from the median are selected. **D** One example of expression profiling done on a set of known genes, using clustering software developed by Michael Eisen, Stanford (http://rana.stanford.edu/software)

time. Since the tendency is to work from ever-decreasing amounts of RNA, the other four parameters that can be tweaked to preserve signal intensities are: (1) the amount of cDNA PCR product spotted on the nylon membrane, theoretically nearly 100 ng can be fixed for a spot of 200 μm diameter), (2) specific activity of labelled RNA, (3) hybridisation duration, and/or complex probe concentration, and finally (4) exposure time.

A certain percentage of cDNA clones contain repeat sequences, and give a detectable signal with Cot1 DNA. Annealing of the probe with Cot1 DNA before hybridisation can attenuate these unspecific signals. However, we prefer to tag these clones as "repeated" and exclude them from further analysis, because they can produce misleading results.

Reverse transcription of the poly(A+) RNA mixture and simultaneous labelling with dCTP-α^{33}P produces many probe molecules beginning with (unlabelled) poly(T) and continuing as specific and labelled mRNA sequences. The abundance of such poly(T)-containing molecules is high, and their hybridisation to unrelated clones via the poly(A) stretch can be a significant problem. This is eliminated by our labelling protocol. Elimination of this artefact is central to the practical use of this system since it could generate spurious signals that vary according to the degree of the poly(A) tails, giving rise to false differentials. To control this particular problem, we spot on the membranes, in addition to the other cDNAs, clones containing different stretches of pure polyA tails. Of course, the signal on these control spots must be null in each independent hybridisation with complex probes.

To standardise hybridisation intensities obtained in several experiments, we use an *Arabidopsis thaliana* cDNA sequence *(CG03*, 1-kb insert) which has no homology with mammalian DNA. A defined amount of *CG03* mRNA (spike), transcribed in vitro from the corresponding cDNA clone, is added before labelling to the total RNA of each cell type or tissue to be tested. The quantification of the corresponding *CG03* spots present on each membrane allows each independent hybridisation to be normalised according to the average *CG03* signal; this corrects for differences in the labelling, washing, duration of exposure as well as progressive degradation of the membranes (Fig. 4.6; Bernard et al. 1996). The exact amount of this exogenous RNA spike must be chosen to fall within the interval of moderately expressed species of the RNA population. Experimental variations can be taken into account by measuring this external control, so that differential expression levels quantified for each clone between any independent hybridisations can be compared with greater confidence. Fluorescence analysis, in contrast, is done in relative probe-to-probe mode (thus avoiding most of the experimental variations in differential analysis between the two probes); however, beyond two probes, a common reference RNA is needed to perform meaningful comparisons.

5 Protocol 1: Array Preparation

5.1 Equipment and Reagents

- Arrayer robot workstation BIOGRID with a 96- or 384-pin tool (Bio Robotic).
- GMS 417 Arrayer (Genetic MicroSystems, Affymetrix), pin and ring tool.
- Thermal Cycler 9700 (Perkin Elmer).
- Thermowell (thin-wall 96-well plate), poly labo ref 67962 and 66867
- 96-well plates (V bottom) (COSTAR).

1. LB: 10 g/l bacto-trypton, 5 g/l bacto-yeast extract, 5 g/l NaCl adjusted to pH 7, and LB agar for plates: LB and bacto-agar 1.5% (w/v). Autoclaved.
2. Macroarrays: 8×12 cm nylon filters (HYBOND-N+, Amersham).
   Microarrays: 2.5×7.6 cm nylon filters (HYBOND-N+, Amersham)
3. 3MM papers 20×20 cm placed inside a 22×22 cm Nunc plate.
4. Spray rub 3 M.
5. Denaturing solution: 0.5 M NaOH, 1.5 M NaCl freshly prepared.
6. Neutralising solution: 1 M Tris-HCl, pH 7.4, 1.5 M NaCl
7. Deproteinisation buffer: 50 mM Tris-HCl, pH 7.4, 50 mM EDTA, 100 mM NaCl, 1% Na-lauryl-sarcosine (w/v) containing 250 μg/ml of proteinase K freshly prepared.
8. PCR primers.
   Primary PCR primers: Tm 58 °C
   LBP 1S: Tm 66 °C 5′ GTG GAA TTG TGA GCG GAT AAC 3′
   LBP 1AS: Tm 58 °C 5′ GCA AGG CGA TTA AGT TGG G 3′
   Common vector primer:
   LBP 3: Tm 54 °C 5′ TGT AAA ACG ACG GCC AGT G 3′
   Original primers:
   T3: Tm 46 °C 5′ ATT AAC CCT CAC TAA AG 3′
   T7: Tm 46 °C 5′ AAT ACG ACT CAC TAT AG 3′
   T7bis: Tm 74 °C 5′ GGG AAT TTG GCC CTC GAG GCC AA 3′
   Oligonucleotide 15: Tm 58 °C 5′ TGT GGA ATT GTG AGC GGA TA 3′
   Oligonucleotide 16: Tm 60 °C 5′ GTT TTC CCA GTC ACG ACG TT 3′
9. 10× PCR buffer (Promega).

5.2 Protocol for Colony Membranes

Adaptation of protocol described by Nizetic et al. (1992).

1. Colonies from freshly grown replica plates are spotted on to 8×12 cm membranes placed on the top of LB agar plates (containing antibiotic).

2. Each colony is spotted following a pattern already designed and stored in computer.
3. Plates carrying membranes are placed bottom up and incubated at 37 °C for approximately 12 h (colony sizes are checked after 10 h of growth. Size should be 0.5–1 mm.
4. Denaturation is performed by carefully placing the membranes, colony side up, on 3MM paper soaked with 50 ml of denaturing solution for 4 min at room temperature followed by a second treatment in the same buffer for 4 min at 80 °C in a damp atmosphere.
5. Repeat this step twice.
6. Membranes are then neutralised by placing them successively, on 3MM papers soaked with 50 ml of neutralising solution for 4 min each at room temperature.
7. Repeat this step twice.
8. Protein is then removed by treating the membranes with 50 ml of deproteinisation buffer for at least 2 h at 37 °C.
9. Membranes are rinsed one by one in 100–200 ml of 2× SSC.
10. Air dry on paper (never pile membranes until they are completely dry, otherwise cDNA can stick to the back of the above paper, losing part of the material).
11. The DNA is fixed by treatment at 80 °C for 2 h followed by UV cross-linking (230 nm, 0.16 KJ/m^2).
12. Arrays are ready and can be stored at room temperature.

5.3 Protocol for PCR Membranes

Rather than colonies, PCR products from the same cDNA libraries can be spotted on membranes. In this case spotting condition and treatment are slightly different; macroarrays and microarrays can be made depending on the arrayer workstation.

5.3.1 Sample Preparation

1. Each individual bacterium corresponding to the set of clones selected to be spotted is grown at 37 °C OVN, in 100 µl of LB plus antibiotic in 96-well plates (V bottom).
2. Pellet bacteria in 96-well plates by centrifugation for 5 min at 1,500 rpm (Jouan GR 412).
3. Remove the supernatant and add 100 µl H_2O.
4. Repeat the centrifugation (10 min, 1,500 rpm).

5. Remove the supernatant.
6. Suspend pellet in 100 µl H_2O.
7. At this step bacteria are ready for PCR reaction and can be stored at -20 °C for a while.

5.3.2 PCR Amplification

For each clone, between 2 and 4 PCR are performed, pooled and concentrated by evaporation before being resuspended in 20–40 µl of water in order to reach a DNA concentration between 200 and 500 ng/µl.

1. PCR reactions are prepared by filling thermowell with 90 µl of a mix-PCR/well.
2. 10 µl of each bacterial suspension is added and mixed.
3 The PCR reaction is performed in 100 µl.

Mix-PCR/reaction:	70.5 µl H_2O
	10 µl 10× PCR buffer
	6 µl $MgCl_2$ (25 mM)
	1 µl mix primers (100 µM)
	2 µl mix dNTP (10 µM)
	0.5 µl (Taq Polymerase)
Denaturation:	94 °C, 6 min
Cycle: 40-fold	94 °C, 30 s
	55 °C, 40 s
	72 °C, 1 min
Elongation:	72 °C, 10 min

4. PCR products are tested on a 1% agarose gel.
5. Pooling and evaporation of individual PCR products are performed sequentially: when the 96 individual PCR products from the first plate are evaporated on the thermocycler (94 °C), the PCR products of the second copy plate are added to be evaporated, and so on until the end.
6. Final pellets are resuspended in 30 µl of distilled water.
7. At this step plates can be stored at 4 °C or at -20 °C until spotting.

5.3.3 Spotting

1. Stick precut nylon membrane with a small amount of spray rub: on glass slide of 2.5×7.6 cm^2 for microarrays, or place on 8×12 cm^2 Whatman paper for macroarrays.
2. Organise nylon slides on the robot tray.
3. Hygrometry should be controlled and stabilised at 65–75% to avoid sample evaporation during the spotting.

4. Start the spotting, the number of hits for each spot depending on the concentration of the PCR product and on the pin design, which is different between robots (see Table 4.1).

5.3.4 Membrane Treatment

This step is much easier than with colony membranes since the DNA is directly spotted.

1. Remove delicately the membrane from the support with the help of forceps.
2. Denaturation is performed by carefully placing the membranes, DNA side up, on 3MM paper soaked with 50 ml of denaturing solution for 10 min at room temperature.
3. Repeat this step twice.
4. Neutralise membranes by placing them on 3MM papers soaked with 50 ml of neutralising solution for 10 min each at room temperature.
5. Repeat this step twice.
6. Rinse membrane by baking them into 100–200 ml of 2× SSC.
7. Arrays are then dried at room temperature without dust.
8. DNA is fixed by treatment at 80 °C for 2 h followed by UV cross-linking (230 nm, 0.16 KJ/m^2).
9. Arrays are ready and can be stored at room temperature.

5.3.5 Notes

1. Up to 3,000 colonies can be spotted on an 8×12 cm^2 macroarray which means an 8×384 well plate.
2. Pitch for macroarrays is at least 2 mm for colony membranes because of bacterial growth. This inter-space can be reduced for PCR spotting; the limitation then depends solely on the image acquisition system (Fuji BAS 1500, 100-μm pixel size). Interspot spacing for microarrays can be as low as to 375 μm, taking into account the image acquisition system limitations (Fuji BAS 5000, 25-μm pixel size, Biospace, 5 μm) as well as the intrinsic limits due to radioactive labels.
3. Colony membranes are stored dry in clean plastic bags at room temperature until the first use. Then when membranes have been hybridised, they can be kept (after stripping) submerged in the strip buffer at 4 °C, or wet in a plastic bag at –20 °C.

6 Protocol 2: RNA Preparation

When working with RNA, one should not use any plastic- or glassware without first eliminating possible ribonuclease contamination (unless it is disposable and individually wrapped by the manufacturer). Only sterile, new pipette tips and microfuge tubes should be used, and clean microbiological aseptic techniques performed. For further information on controlling RNase contamination see Sambrook et al. (1989).

6.1 Equipment and Reagents

1. "Trizol" reagent (GIBCO-BRL, Bethesda, MD).
2. DEPC water: 0.1% diethyl pyrocarbonate in water, mixed overnight then autoclaved to remove traces of DEPC that might otherwise modify purine residues in RNA by carboxy methylation.

6.2 RNA Extraction

1. Total RNA is isolated from cell lines following the instruction manual provided with the "Trizol" kit (GIBCO-BRL, Bethesda, MD). Suspending RNA in DEPC water under standard conditions, we obtain 10 µg of total RNA per million cells.
2. The RNA messenger of *Arabidopsis thaliana* chlorophyll synthetase was obtained from cDNA cloned into Bluescript SK+ vector at the Not I restriction site, and was synthesised from the T3 promoter using the RiboMax large-scale production system (Promega).

Caution: DEPC is suspected to be a carcinogen and should be handled with care under a hood.

7 Protocol 3: Labelling

7.1 Equipment and Reagents

1. RT buffer mix: 1 µl of RNAsin (RNase inhibitor, Promega, ref. N2511, 40 U/µl), 6 µl of 5× first strand buffer (BRL), 2 µl of 0.1 M DTT (BRL), 1 µl (0.6 µl) of dATP dTTP dGTP mix (20 mM each), 1 µl (0.6 µl) of 120 µM dCTP, 5 µl (3 µl) of 10 µCi/µl (alpha-33P) dCTP (>3,000 Ci/mM), 2.8 µl H_2O, 1 µl of reverse transcriptase (superscript RNase H free RT, BRL, 200 U/µl).
2. Oligonucleotide dT25.
3. Oligonucleotide dA80: 3′-$(dATP)_{80}$-5.′
4. According to sequence on use, 15- to 20-mer unphosphorylated oligonucleotides are selected: for 28S, 5′ TGAATCCTCCGGGCGGACT (for Northern blots); for other vectors (LBP 3), 5′ TGACCGGCAGCAAAATGT.
5. Ready Cap scintillating capsule (BECKMAN).
6. Sephadex G50 or G25 column: Sephadex G50 (Pharmacia) is swelled in water and autoclaved. The column is prepared in a 1-ml syringe plugged with glass fibre (Whatman GF/B). Correct placement of the plug is carefully observed to avoid recovering pieces of gel in the probe, which would provide spurious hybridisation spots. The column is centrifuged repeatedly for 2 min at 1,000 rpm (Jouan GR 412) after adding 150 µl H_2O, until centrifugation yields only 150 µl of liquid (to pack column). The column is ready when the remaining G50 is between 0.9 and 1 ml (after two to three rounds). At the last run when the liquid is removed, an Eppendorf tube is placed at the bottom of the column and the probes (150 µl) are loaded, and recovered by centrifugation.

7.2 Oligonucleotide Labelling

The vector oligonucleotide is labelled with [gamma-P^{33}] ATP at the 5′ end using standard methods (31) briefly described below.

1. Mix 1 µl oligonucleotide (1 µg/µl), 2 µl 10× T4 kinase buffer (Biolabs), 3 µl of [gamma-P^{33}] ATP (3,000 Ci/mM), 1 µl of T4 polynucleotide kinase (10U/ml, Biolabs) and complete to 20 µl with water.
2. Incubate for 45 min at 37 °C.
3. Eliminate most unincorporated ATP by G25 column (see above) or precipitation:
   - Add: 1 µl herring sperm DNA (Boehringer, 10 mg/ml), 2 µl 3 M sodium acetate, 60 µl absolute ethanol.

- Place 15 min at –80 °C.
- Spin 30 min at 4 °C.
- Discard supernatant *(highly radioactive!!!)*.
- Resuspend pellet in 100 µl sterile water.

4. Count 1 µl, usually 30–50 million cpm for the whole probe.

Note: the unincorporated nucleotides can be removed by purification on a Sephadex G25 column (see above).

7.3 Complex Probe Labelling from Total RNA

Annealing. To remove secondary structure of RNA and saturate polyA tail with oligo-dT. A large excess of oligo-dT is used so that later RT transcription will start near the beginning of the polyA tail.

1. Mix 25 µg (5 µg) total RNA in 11 µl DEPC water with 2 ng (0.4 ng) CG03 chlorophyll synthetase mRNA at 0.5 ng/µl (4 µl) and 8 µg of dT12–18 in an Eppendorf tube.
2. Place sample for 8 min at 70 °C in a water bath to remove secondary structure.
3. Progressively cool the mixture to reach 42 °C. This is performed by placing the tube in a metal block preheated at 70 °C, then the block is backed into an oven set at 42 °C. This step should take 30 min.

Note: usually when several complex probes must be made at the same time, a mix containing the CGO3 mRNA, oligo-dT25 and water is prepared for the whole series.

Reverse transcriptase. To simultaneously synthesise and label single-stranded DNA from 500 ng (or 100 ng) of mRNA present into 25 µg (5 µg) of total RNA.

1. Add mix RT buffer to the tube and keep at 42 °C (in the block).
2. Incubate the reaction for 1 h at 42 °C in the oven.
3. Add 1 µl of enzyme and incubate for an additional hour.

Removal of RNA. To degrade mRNA and rRNA and obtain a single-stranded probe.

1. Remove the RNA by treatment at 68 °C during 30 min with 1 µl of 10% SDS, 1 µl of 0.5 M EDTA, and 3 µl of 3 M NaOH.
2. Equilibrate the complex probe at room temperature for 15 min.
3. Neutralise the probe by adding 10 µl of 1 M Tris pH 7and 3 µl of 2 N HCl.

Purification of complex probe. The unincorporated nucleotides are removed from the complex probe by purification on a Sephadex G50 column, otherwise they increase background noise.

1. Load probe (150 µl) on top of the column.
2. Spin for 4 min at 1,000 rpm.
3. Recover 1 µl out of the 150 µl after spinning and count. Normally, the total radioactivity in the probe is around 30 million counts using Ready Cap scintillating capsules.
4. Block the possible polyT tail left during the first treatment by adding 2 µl dA80 solution at 1 µg/µl to the probe and Repeat sequences by adding 10 µl of cot 1 (1 µg/µl) (Gibco. BRL).
5. Denature the whole probe for 5 min at 100 °C.
6. Add 1 ml of hybridisation buffer (preheated to 65 °C) to the probe.
7. Incubate for 2.5 h at 65 °C before adding to the 50 ml of hybridisation buffer.

8 Protocol 4: Hybridisation

8.1 Equipment and Reagents

1. Water bath, 65 °C.
2. Rotisserie oven (Appligène, France)
3. 20× SSC: 175.3 g of NaCl and 88.2 g of sodium citrate is dissolved in 800 ml of distilled H_2O. The pH is adjusted to 7, completed to 1 l and the solution is autoclaved.
4. 10% SDS (W/V) in distilled sterile H_2O.
5. Denhardt's reagent 100×: 10 g Ficoll (Type 400, Pharmacia), 10 g of polyvinylpyrrolidone, 10 g of bovine serum albumin (fraction V, Sigma), are dissolved in distilled sterile H_2O, then completed to 500 ml.
6. Hybridisation buffer: 5× SSC, 5× Denhardt's mix, 0.5% SDS. Sonicated and denatured herring sperm DNA is added just before hybridisation (final concentration: 100 µg/ml). For microarrays, buffer is filtered on 0.8-mm Millipore membrane before being used.
7. The herring sperm DNA (Boehringer) is stored sonicated at 10 mg/ml in aliquot of 1 ml. Denaturation is performed just before being used by heating 10 min at 100 °C, then cooling rapidly to 0 °C by placing the tube for 10 min on ice.

8.2 Hybridisation with an Oligonucleotide: "Vector" Probe

This hybridisation is performed in order to quantify the cDNA fixed onto the membranes and should be done under saturating conditions. For this purpose unlabelled oligonucleotide is added before the hybridisation. If this is omitted, specific activity is very high and irreversible adsorption of a very small amount of very highly labelled oligonucleotide will be a problem: the signals will not disappear on stripping. Adding unlabelled oligonucleotide avoids this problem and also ensures high probe concentration that leads to good signals in spite of lower specific activity, and reliable measurement conditions are close to saturation.

1. Pre-hybridise at least 4 h in hybridisation buffer (50 ml and 4 membranes per box). Microarrays are hybridised in small Wheaton tubes with 1 ml of buffer.
2. Hybridisation is performed in a pre-hybridisation buffer; remove membranes before adding the probe.
3. Add an amount of probe corresponding to 100,000 to 200,000 cpm/ml and at the same time 10 µg of unlabelled oligonucleotide.
4. Place the membranes back one by one and hybridise at least 12 h at 42 °C in a water bath with shaking (along the small side of membranes).
5. Wash in large excess (up to 500 ml) of 2× SSC, 0.1% SDS for 10 min at room temperature, then change buffer and wash 5 min at 42 °C.
   Expose wet, plastic-wrapped membrane overnight against imaging plate.

8.3 Hybridisation with Complex Probes

1. Pre-hybridise membranes for at least 6 h at 68 °C with 5× SSC, 5× Denhardt's, 0.5% SDS, and 100 µg/ml sheared denatured salmon sperm.
2. Hybridise with total probes with the same buffer for 48 h. It is unnecessary to change buffer between pre-hybridisation and hybridisation. Best results are obtained when no more than four membranes of 8×12 cm (macroarrays) are hybridised simultaneously per 50 ml in box or one membrane per tube containing 10 ml. Both are performed in a water bath with shaking or tubes. Microarrays are hybridised in small Wheaton tubes with 500 µl to 2 ml of buffer.
3. Wash membranes after hybridisation three times at 68 °C (1 h each) with 1 l of 0.1× SSC, 0.1% SDS. Washing buffer is warmed to 68 °C before use.
4. Expose membranes: wrap in plastic bags (except for microarrays, see below) and place onto phosphor screen for 3 days or more. Plastic bags must be sealed to avoid drying, otherwise a good stripping will not be obtained. Microarrays are exposed dry directly onto the Fuji BAS 5000 imaging plate (high resolution, 25 µm). The distance between the membrane and the imaging plate must be as short as possible to avoid a fuzzy image.

8.4 Dehybridisation

1. Membranes hybridised with an oligonucleotide probe are incubated 3 h at 68 °C in 0.1× SSC, 0.1% SDS. Check by overnight imaging plate exposure and renew dehybridisation if necessary.
2. Membranes hybridised with the cDNA complex probes are stripped twice with the same buffer at 80 °C for 3 to 5 h, then rinsed with 2× SSC. The dehybridisation is controlled by an imaging plate exposure (the exposure time must be as long as the one necessary for complex probes).

8.5 Notes

1. Caution: when handling membranes, wear gloves but do not forget to wash them prior to use. Talcum present on new gloves will give strongly labelled spots on the membranes.
2. Macroarray hybridisation can be performed in 5–10 ml of buffer in a 15-cm-long tube placed in a rotisserie oven (Appligène, France). For microarrays, small Wheaton tubes are blocked into the long tubes before being placed in the rotisserie oven.
3. Washing time is critical for oligonucleotide hybridisation. If you wash too long, even at the correct temperature, you will remove all the oligonucleotide.
4. For 28 S hybridisation (high GC content) a further wash 15 min in 0.2× SSC, 0.1% SDS at 42 °C can be done.
5. Membranes can be re-hybridised at least three times. However, before hybridisation with complex probes, a control must be performed by hybridisation with vector oligonucleotide to assay the amount of cDNA left for each colony on the membrane.

9 Protocol 5: Data Processing

9.1 Equipment

1. Fuji BAS 5000 or 1500 (Fuji) system
2. Bioimage HDG analyser software (Genomic Solutions, USA) running on a SUN workstation.
3. Microsoft Excel software.

9.2 Detection and Quantification of Hybridisation Signals

1. Image acquisition is performed with a Fuji BAS 1500 system. The data occupy a 2-MB file.
2. Import the resulting image files to the quantification software and separate into images of individual membranes.
3. Spots are detected without any assumption of their position. The "quantify" command then determines their shape and intensity individually, with local background subtraction.
4. The tables generated in the workstation are transferred over the network to a microcomputer. A set of Excel macro commands provide an easy conversion of text files into Excel tables in order to perform standard calculations, such as normalisation of data, and produce a standard set of representations used to judge the quality of the data (Fig. 4.6).

9.3 Notes

1. Detection can be performed with a phosphor imager imaging plate system from Molecular Dynamics (Sunnyvale, CA). The resulting 16-bit images are then imported to a Sun workstation to perform the image analysis with the Xdotsreader software (Cose, Le Bourget, France; Pietu et al. 1996); they can also be imported and analysed in the Bioimage software. Image acquisition can be also performed with a Biospace Micro Imager instrument (http://www.biospace.fr) (exposure time 3 days) to analyse nylon microarray with 225-μm pitch. This is a direct radioactivity instrument displaying in real-time quantitative image of radioisotopes deposit on biological samples providing 15-μm spatial resolution with 5-μm pixel size (Bertucci et al. 1999).
2. From a software point of view, the quantification is reproducible. The same image file can be explored using the default settings or the relaxed settings chosen to allow quantification of a maximum number of spots. For the vast majority of spots the procedure is indeed reproducible (Granjeaud et al. 1996).
3. The total signal detected on the membrane as well as the average intensity of the detected spots varies with exposure in a linear fashion. Difference in results due to exposure time or from plate to another are negligible (Granjeaud et al. 1996).
4. Spot finding and spot contouring algorithms are little influenced by spot intensity. For example, vector hybridisation in excess probe conditions yields spot intensities of limited variation which are expected to be correlated with spot size. Indeed, this is what we observed. With the complex probe however, a much wider range of intensities is observed, and, as expected, we see little correlation between spot size and intensity (Granjeaud et al. 1996).

Acknowledgements. We wish to thank all our co-workers who have been instrumental in setting up our hybridisation signature system, as well as Pascal Hingamp and Rebecca Tagett for discussion of the manuscript. Thanks also to Monique Marilley and Jean Thimonier who provided material and permission for Fig. 4.3.

References

Beltz GA, Jacobs KA, Eickbush TH, Cherbas PT, Kafatos FC (1983) Isolation of multigene families and determination of homologies by filter hybridisation methods. Methods Enzymol 100: 266–285

Bernard K, Auphan N, Granjeaud S, Victorero G, Schmitt-Verhulst AM, Jordan BR, Nguyen C (1996) Multiplex messenger assay: simultaneous, quantitative measurement of expression for many genes in the context of T cell activation. Nucleic Acids Res 24:1435–1443

Bertucci F, Bernard K, Loriod B, Chang YC, Granjeaud S, Birnbaum D, Nguyen C, Peck K, Jordan BR (1999) Sensitivity issues in DNA array-based expression measurements and performance of nylon microarrays for small samples. Hum Mol Genet 8(9):1715–1722

Chen JJ, Wu R et al. (1998) Profiling expression patterns and isolating differentially expressed genes by cDNA microarray system with colorimetry detection. Genomics 51(3):313–324

Eisen MB, Spellman PT, Brown PO, Botstein D (1998) Cluster analysis and display of genome-wide expression patterns. Proc Natl Acad Sci USA 95:14863–14868

Granjeaud S, Nguyen C, Rocha D, Luton R, Jordan BR (1996) From hybridisation image to numerical values: a practical, high throughput quantification system for high density filter hybridisations. Genet Anal Biomol Eng 12:151–162

Granjeaud S, Bertucci F, Jordan BR (1999) Expression profiling: DNA arrays in many guises. Bioessays 21:781–790

Gress TM, Hoheisel JD, Lennon GG, Zehetner G, Lehrach H (1992) Hybridisation fingerprinting of high-density cDNA-library arrays with cDNA pools derived from whole tissues. Mamm Genome 3:609–661

Hillier LD, Lennon G, Becker M, Bonaldo MF, Chiapelli B, Chissoe S, Dietrich N, DuBuque T, Favello A, Gish W, Hawkins M, Hultman M, Kucaba T, Lacy M, Le M, Le N, Mardis E, Moore B, Morris M, Parsons J, Prange C, Rifkin L, Rohlfing T, Schellenberg K, Marra M, et al. (1996) Generation and analysis of 280,000 human expressed sequence tags. Genome Res 9:807–828

Imbert MC, Nguyen VK, Granjeaud S, Nguyen C, Jordan BR (1999)'LABNOTE': a laboratory notebook system designed for academic genomics groups. Nucleic Acids Res 27:601–607

Johnston RF, Pickett SC, Barker DL (1990) Autoradiography using storage phosphor technology. Electrophoresis 11:355–360

Jordan BR (1998) Large scale expression measurement by hybridisation methods: from high-density membranes to "DNA chips". Jpn J Biochem (Tokyo) 124:251–258

Mory K, Hamaoka T (1994) IP autoradiography system (BAS). Protein Nucleic Acid Enzyme 39:181–191

Nguyen C, Rocha D, Granjeaud S, Baldit M, Bernard K, Naquet P, Jordan BR (1995) Differential gene expression in the murine thymus assayed by quantitative hybridisation of arrayed cDNA clones. Genomics 29;207–215

Nizetic D, Dramanac R, Lehrach H (1991) An improved bacterial colony lysis procedure enables direct DNA hybridisation using short (10, 11 bases) oligonucleotides to cosmids. Nucleic Acids Res. 19:182

Patterson SD, Latter GI (1993) Evaluation of storage phosphor imaging plating for quantitative analysis of 2-D gels using the Quest II system. Biotechniques 15:1076–1083

Pietu G, Alibert O, Guichard V, Lamy B, Bois F, Leroy E, Mariage-Samson R, Houlgatte R, Soularue P, Auffray C (1996) Novel gene transcripts preferentially expressed in human muscles revealed by quantitative hybridisation of a high density cDNA array. Genome Res 6:492–503

Sambrook J, Fritch EF, Maniatis T (1989) Molecular cloning: a laboratory manual, book 1, 2nd edn. Cold Spring Harbor Laboratory Press, Cold Spring Harbor, pp 7.3–7.5

Supplement to Nature Genetics (1999) The chipping forecast. January, vol 21

CHAPTER 5

Oligonucleotide Chips for Expression Analysis: Principles and Practical Procedures

Pierre Casellas[1], Annick Peleraux[1] and Sylvaine Galiegue[1]

[1] Immunology-Oncology Department, Sanofi-Synthelabo,
371 Rue du Professeur Joseph Blayac, 34184 Montpellier, Cedex 04, France

1 Introduction

The analysis of biological processes has been revolutionized by the emergence of DNA array technology. In the cell, nucleic acid-related phenomena are intricate and complex, but now they can be analyzed globally. As cellular biological events are controlled by gene expression, their modulations are markers of cellular activity and their analysis can be highly informative. These modulations may be specific for a given cell type and/or regulated over time. They can be indicative of either a physiological process or a pathological one. Monitoring of the expression levels of thousands of genes simultaneously, the expression profiling method described here, is based upon comparative studies where the identification of differentially expressed genes in two samples is aimed. The two samples under study may be compared temporally or following drug treatment; they may also originate from different sources, e.g. normal versus pathological samples. In that case, gene expression profiling is conducted for diagnostics purposes or therapy monitoring, and offers an opportunity to identify new drug targets.

Here we focus on the oligonucleotide chip technology. Illustrating with an example, we describe the potential impact of this technology for measuring gene expression profiles. At the end of the chapter, we also highlight the need for bioinformatics resources for the analysis of the data obtained.

2 Principles

2.1 Affymetrix DNA Chip Design

We focus on oligonucleotide arrays designed by Affymetrix, a technology commonly used in our laboratory since 1997. Oligonucleotides are synthesized based on sequence information available in the public databases. For Affymetrix chip design, the manufacturer preferably uses the 3′ region to represent a gene. 16- to 25-mer oligonucleotide sequences for each gene of interest are synthesized onto the array. The oligonucleotide probes are organized as perfect match/mismatch pairs, with the mismatch probe differing from the perfect match probe by only one base. The mismatch probe acts as a control for hybridization specificity. A gene is then represented by 40 oligonucleotides: 20 perfect match and 20 mismatch. The hybridization unit where millions copies of a nucleotide are synthesized (either match or mismatch probe) is the main feature of the chip. The hybridization unit is commonly a 24×24 μm square and the reduction of its size is aimed to increase the density of genes per chip. Oligonucleotide synthesis takes place on glass using a light-directed, solid-phase combinatorial chemistry approach that relies on photolithographic fabrication techniques (McGall et al. 1996). High-density oligonucleotide arrays are synthesized using photolithographic masks that define chip exposure sites where specific chemical synthesis takes place after light deprotection. Multiple probe arrays are synthesized simultaneously and individual arrays are packaged in protective injection-molded plastic cartridges that serve as chambers for hybridization. Chips are provided as sets of 1 to 5 arrays. They have to be used only once. The shelf life of the Affymetrix array is 6 months from the date of manufacture when stored at 2–8 °C.

The available Affymetrix probe arrays are designed from different sources and in different sets: human (U95 set, 5 arrays for the analysis of >60,000 full-length genes and EST clusters; Hu35 K set, 4 arrays providing gene expression data for 35,000 EST sequences from the Unigene database; HuGene FL array for the analysis of 6,800 full-length sequences on a chip), rat (U34 set, 3 arrays for the analysis of >24,000 known genes and EST clusters), mouse (Mu11 K, 2 arrays for the analysis of 30,000 murine genes and EST clusters), yeast (S98 array for the analysis of the 6,400 ORF of the yeast *Saccharomyces cerevisiae*), *Escherichia coli* (for the analysis of 4,200 ORF), *Arabidopsis thaliana* (for the analysis of 8,200 genes and EST). Other arrays are available for specific purposes (Human cancer, HuSNP, Rat toxicology, Rat neurobiology, HIV, p53 or CYP450 arrays) or can be made upon request.

Detailed information on the Affymetrix technology and products can be obtained from the web site: http://www.affymetrix.com.

2.2 DNA Chip Procedure

Gene expression analysis is a combination of several interrelated steps: isolation of RNA, reverse transcription to cDNA, in vitro transcription and biotinylation, fragmentation, hybridization to the array, fluorescent labeling, and optical scanning of the fluorescent chip. The sample preparation takes approximately 2 days. The mRNA is first isolated from the target cells or tissue. The preferred quantity of mRNA is about 1 µg, although less is needed if single chip analysis is required. As typical yield is 1–5 µg mRNA from 100 µg total RNA, the sample preparation may include a preliminary step to adjust the cell number or the sample size to the experiment. mRNA is reverse transcribed to double-strand cDNA using conventional protocols. The double-strand cDNA is transcribed to cRNA and at the same time is biotinylated using biotin-CTP and biotin-UTP. The combination of reverse transcription and in vitro transcription is a round of amplification that produces 60–100 µg labeled cRNA from 1–2 µg mRNA. The biotinylated cRNA is then chemically fragmented (Fig. 5.1A), mixed with control cRNA fragments, and hybridized to an oligonucleotide array; 10–20 µg of labeled cRNA is hybridized to a single chip. Hy-

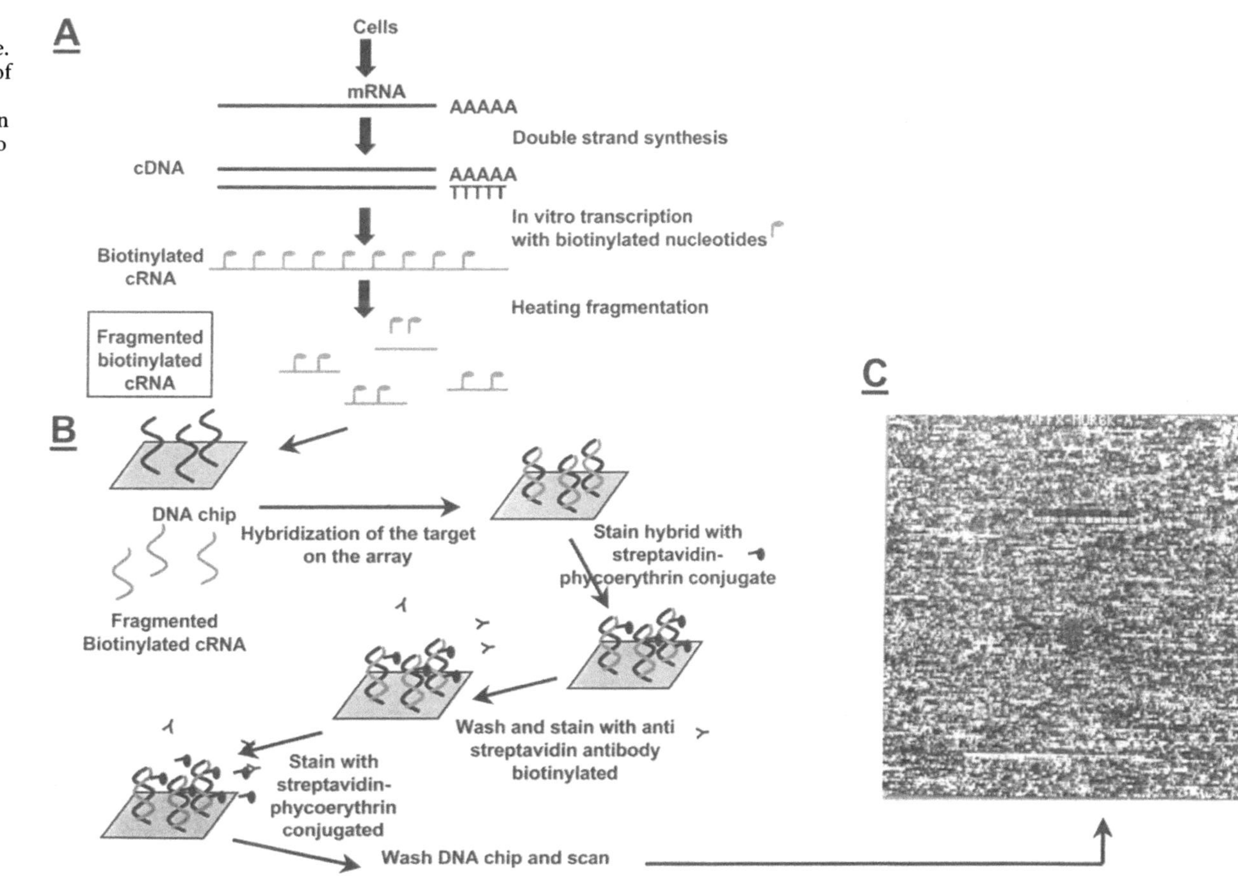

Fig. 5.1. DNA chip procedure. **A** Preparation of RNA samples. **B** Hybridization of samples onto the chip and labeling step. **C** Example of a scan

bridization is usually performed overnight (16 h). The hybridized biotinylated cRNA is fluorescently labeled by strepdavidin-phycoerythrin. The signal is enhanced by a two-step labeling (Fig. 5.1B). Washing and staining take approximately 1 h and 15 min respectively. After washing, the oligonucleotide chip is scanned for 10 min to provide data for functional analysis (Fig. 5.1C).

2.3 Analysis of Biological Process: Example of Analysis of CB2 Signaling

Enabling the analysis of complex expression profiles, the oligonucleotide chip technology has opened up new opportunities in disciplines ranging from cell and developmental biology to drug development and pharmacogenomics. The use of expression profiling in pharmacological studies is here exemplified by the DNA chip analysis of the signaling pathway originated from the binding of a ligand to its receptor: in this case the peripheral cannabinoid receptor activated by the binding of its ligand CP 55,940.

We describe the experimental procedures, the definition of the statistical parameters for the validation of the analysis and illustrate the biological meaning extracted from the DNA chip data.

2.3.1 Biological Context

Two cannabinoid receptors have been described and are referred to as CB1 and CB2 (Devane et al. 1988; Matsuda et al. 1990; Gérard et al. 1991; Munro et al. 1993). Since their identification, they have been the objects of intensive research in order to understand the wide spectrum of central and peripheral effects induced by cannabinoids. The central receptor CB1 is predominantly expressed in the brain whereas the peripheral receptor CB2 is mainly expressed on cells of immune origin (Kaminski et al. 1992; Bouaboula et al. 1993; Galiègue et al. 1995). Both are G-protein coupled receptors (Childers et al. 1993). Activation of CB1 has been shown to lead to the inhibition of adenylyl cyclase and the inhibition of N-type voltage-dependent calcium channels both in vitro and in vivo (Caulfield and Brown 1992; Schatz et al. 1992). The biological effects induced by cannabinoids include alterations in cognition and memory, analgesia, anticonvulsion, alleviation of both intraocular pressure and emesis (Howlett 1995). In addition to these central effects mediated by the CB1, many reports have also described a cannabinoid-induced modulation of immune functions that could be attributed to the peripheral receptor subtype, CB2. With the exception of few examples reporting stimulating effects, most of these studies showed immunosuppressive properties of cannabinoids in very different areas of the immunity, including humoral and cellular responses as well as cytokine production. To date, the function of CB2 receptors in the immune system is still elusive. We describe here the use of oligonucleotide chips and expression profiling to decipher the signaling events associated with the activation of CB2 receptors following ligand binding. We have indeed investigated the effect of nanomolar concentrations of a specific cannabinoid ligand (CP 55,940) on the transcriptional program induced in the promyelocytic cell line HL-60.

2.3.2 Experimental Procedures

2.3.2.1 Sample Treatment and RNA Extraction

In this example, the cells used were the HL-60 promyelocytic cells transfected with the CB2 receptor. This cell line, lacking central CB1 receptors, made it possible to focus specifically on the peripheral receptor subtype CB2 which was further overexpressed by transfection in order to amplify a possible cannabinoid-induced effect on gene modulation. Cells were grown in culture medium with 0.5% fetal calf serum (FCS) 24-h prior treatment with 10 nM CP-55, 940 or 200 nM SR 144528 for different time periods. RNA is first isolated from 10^8 cells using the guanidinium isothiocyanate method (Chomzcynski and Sacchi 1987). Poly A+ RNA was isolated using Fast Track 2.0 Kit (Invitrogen NV, Leek, The Netherlands).

2.3.2.2 cDNA Synthesis

Double-stranded cDNA was prepared from 1.5 µg polyA RNA using Life Technologies superscript choice system and 100 pmol of an oligo$(dT)_{24}$ anchored T7 primer.

2.3.2.3 In Vitro Transcription

Biotinylated RNA was synthesized using the T7 Megascript system kit (Ambion, Texas) with biotin-11-CTP and biotin-16-UTP for 5 h at 37 °C. In vitro transcription (IVT) products were purified using microspin S200 HR columns (Amersham-Pharmacia LKB, Orsay, France).

2.3.2.4
Fragmentation

Twenty micrograms of biotinylated cRNA was then treated for 35 min at 94 °C in a buffer composed of 200 mM tris-acetate, pH 8.1, 500 mM potassium-acetate, and 150 mM magnesium-acetate.

2.3.2.5
Hybridization, Washing and Scanning

Affymetrix huGene FL array was hybridized with biotinylated IVT products (10 µg/chip) for 16 h at 45 °C using the manufacturer's hybridization buffer. The fluidic station 400 from Affymetrix was used for washing and staining the arrays according to an automated procedure. Due to the size-reduced hybridization features for the huGene FL array (24×24 µM), a three-step protocol was used to enhance the detection of the hybridized biotinylated RNA. It consisted in a first incubation with a streptavidin-phycoerythrin conjugate followed by a labeling with an anti-streptavidin goat biotinylated antibody (Vector Laboratories, Burlingam, CA) and a final staining again with the streptavidin-phycoerythrin conjugate. The DNA chips were then scanned using a specific scanner (Hewlett-Packard). The excitation source was an argon ion laser and the emission was detected by a photomultiplier through a 570-nm-long pass filter. Digitized image data were processed using the GeneChip software (version 3.1) available from Affymetrix.

2.3.3
Critical Aspects for Validation of the Analysis: Definition of Statistical Parameters

In order to validate the gene expression modulation observed using the DNA chip technology one must first be aware that a robust statistical analysis of different hybridization parameters is necessary (Cohen et al. 2000).

2.3.3.1
Reproducibility

First, to examine the reproducibility of hybridization signals obtained by the oligonucleotide DNA chips, the same complex probe sample is hybridized to two different sets of arrays, and a good reproducibility between repeated measurements (r=0.99) is found. The hybridization of two samples independently prepared from the same samples enables us to assess differences due to the preparation of the complex probe rather than the hybridization and reading steps. In that case a good correlation was observed (r=0.96), although not as good as the one obtained in the simple hybridization reproducibility experiment: this might reflect different handling and separate in vitro transcription reactions. In large-scale gene expression measurements, sensitivity of mRNA detection is a crucial parameter, because most of the mRNAs are present as a few copies per cell. The general pattern of distribution of the signals obtained with the DNA chips technology was in agreement with typical figures of mRNA population distribution, as the majority of the signals were skewed towards the low intensity values.

2.3.3.2
Fold Change

The fold change (FC) parameter is the most indicative one for comparing gene expression modulation between two samples. In usual large-scale measurement studies, FCs above a two-, three- and sometimes four-fold variation have been considered as significant, as it was logical that the lower an FC was, the less reproducible it was. Nevertheless, we may note that FCs as low as 1.8 could be confirmed by Northern blotting experiments (data not shown). An aspect of the GeneChip software is to allow the identification of regulated genes whose expression was undetectable in one of the two tested samples. In that case, the FC was calculated by dividing the hybridization intensity value for the gene in the sample where it was detected as present by a minimal value of 20, which approximates the limit of detection; thus the calculated FC most likely represents an underestimation of the true FC. Finally, we frequently observed in the reproducibility experiments that while the absolute assignment of a given gene (increased or decreased) was reproducible, the absolute value of the FC might vary from experiment to experiment, and therefore it might be rejected from the analysis if too stringent criteria were applied.

2.3.3.3
Signal Intensity

In the overall analysis, in addition to the FC, the intensity signal has also to be considered. We observed that intensity values of less than 60 are less reliable than high intensity values and should be cautiously considered when a variation of gene expression is detected.

2.3.4
Deciphering CB2 Signaling Using Gene Expression Profiling

Oligonucleotide chip technology is particularly well suited for analyzing the gene expression profile induced by the activation of receptors whose function is still poorly understood. The identification of the genes up- or downregulated

Table 5.1. Cannabinoid-induced modulation of genes in CB2-HL-60. Biotinylated RNA from cells treated with 10 nM CP 55,940 for 1 h, or untreated, were hybridized onto Affymetrix huGene DNA chips and quantified. Fold change of modulated genes is indicated. Genes are classified according to their known function

Genes involved in cytokine production and regulation	Accession no.	Fold change
IL-8	M28130	+2.3
MCP-1	HG4069-HT4339	+2.8
MIP-1β	M69203	+4.7
TNF-α	X02910	+2.8
A20	M59465	+2.2
Genes involved in transcription process and cell cycling	**Accession no.**	**Fold change**
jun-B	X51345	+3.5
Aldolase C	X05196	+3.4
BTG2	U72649	+3.5
IκB-α (MAD3)	M69043	+3.9
Miscellaneous	**Accession no.**	**Fold change**
Metallothionein	V00594	–1.7

following CB2-receptor stimulation was performed by pairwise comparison of the data obtained from computer processing and analysis of the oligonucleotide array hybridized with the sample treated with 10 nM CP 55,940 for 1 h with those obtained from the control array. Ten genes were modulated after stimulation of the CB2 receptor and displayed greater than a two-fold change in their expression level. These genomic modulations were reproducibly observed in three different experiments (Table 5.1).

The first group of genes shown to be upregulated encodes inflammatory mediators and a related regulating factor. Genes encoding three chemokines (MCP-1, IL-8 and MIP-1β) and one cytokine (TNF-α) were consistently upregulated. Further, the TNF-α-responsive gene A20 was also shown to be upregulated. It encodes a zinc finger protein whose expression protects cells from TNF-α cytotoxicity by interacting with TRAF2, one of the adaptor proteins that associate with the cytoplasmic death domain of the TNF-receptor superfamily. Its expression correlates with the lymphocyte activation and monocyte differentiation and is under the control of the transcription factor NF-κB. A second major set of strongly upregulated genes (fold change >3) was the group of genes encoding proteins involved in the transcription machinery and cell cycling. The immediate early gene Jun-B is a member of the transcription factor complex AP-1. Like other products of jun and fos genes, it plays an important role in the transcription of genes involved in cell cycling and differentiation processes. Aldolase C is one of the four glycolytic enzymes described so far with dual functions. The first is a role in the glycolytic pathway and the second is related to their DNA-binding properties. Thus, aldolase C is reported to take part as a nuclear factor in the stabilization of DNA synthesis and to be involved in the transcription of genes associated with cell growth and differentiation. BTG2 is an antiproliferative p53-dependent component of the DNA damage cellular pathway with a major role in growth control and differentiation. The gene IκB-α (MAD-3), which displayed an almost four-fold enhancement, encodes a cytoplasmic protein which regulates the activity of the ubiquitous transcription factor NF-κB. Finally, metallothionein was the only gene shown to be downregulated. It encodes a metal-binding protein involved in the defense against heavy metal-induced injury.

A crucial step in global analysis of the expression modulation is to corroborate the transcriptional changes with protein modulation, as the proteins are the effectors of the biological processes. The modulations described here have been assessed at the protein level (data not shown). The receptor specificity of this response has also been demonstrated by the blockade of the induced effects by the CB2 receptor-antagonist SR 144528 (Rinaldi-Carmona et al. 1998).

The biological meanings of the modulations observed following the CB2 activation are two-fold. First, the stimulation of the CB2 receptor results in the activation of an NF-κB pathway. This is reinforced by the fact that 70% of the genes shown here to be upregulated are under the control of this transcription factor. The pleiotropic transcription factor NF-κB plays a major role in gene regulation during cell differentiation and immuno-inflammatory reactions. Most of the genes evidenced as modulated following CB2 activation are reported in the literature to be associated with myeloid cell differentiation. Secondly, the transcription program elicited by the activation of the peripheral CB receptor indicated a possible role for this receptor in the initialization of a cell maturation process. HL-60 cells undergoing maturation under the effects of differentiating agents are known to display cytoskeletal rearrangements underlying cell migration (Cross et al. 1997). Our data indicate that CP 55,940 induced an increase in the cell migration that was enhanced cell motility rather than true directional migration, suggesting that CP 55,940 was chemokinetic rather than chemotactic (data not shown). A possible action of the cannabinoid agonist is to trigger mechanisms of cell locomotion that could further facilitate the subsequent action of true chemotactic agents. Activation through the peripheral cannabinoid receptors is here shown to induce in HL-60 cells an intermediate stage of maturation evidenced by the genomic upregulation of factors involved in cell differentiation, an enhanced synthesis of chemotactic proteins and an increased locomotion which are key events in leukocyte trafficking and the inflammatory response. Altogether these results suggest that CB2-receptor activation could induce conditions that facilitate the transition of HL-60 cells to a more mature monocytic/granulocytic phenotype.

This study evidences the use of oligonucleotide chips technology in the functional analysis of a receptor: expression profiling provides crucial information that opens up new opportunities in the understanding of the CB2-receptor activity.

3 Bioinformatics

A major aspect of DNA array technology is the absolute requirement for bioinformatics including data storage and organization, data analysis and interpretation. Due to the sheer amount of data generated, all these steps have to be carefully designed and automated for the most part.

3.1 Data Management

One must first be aware of the amount of memory space required for data storage. For example, 60 MBytes is necessary for storing data files documenting the expression of 6,000 genes in one sample using the Affymetrix technology. Also, several files are produced for each DNA chip analyzed, namely an image file (43 MBytes), an experiment description file, a text file describing the average intensity of fluorescence of each hybridization unit, and an analysis file containing all parameters calculated for each gene. An additional analysis file is created for each comparison between two chips, hybridized for example with treated and control sample cRNA. All these files need to be named and organized in a systematic way if one wants to easily relate them to the corresponding chips, samples and experiments. Despite such precautions, data consultation rapidly becomes an intricate task when several series of results have to be viewed at once, if the data are kept in individual files. To circumvent this problem, data can be organized in databases, allowing queries pertaining to many different chips, samples or experiments. Grouping all the data in a secure database also eliminates the risk of data losses, and allows simultaneous treatment of data from any experiments. Affymetrix provides a Laboratory Information Managing System (LIMS) able to store the relations between the files, and to publish the analysis results in a normalized relational database designed to store all parameters calculated by the GeneChip software. The specifications of the database were defined by the Genetic Analysis Technology Consortium and are available on the Internet (http://www.gatconsortium.org).

3.2 Data Analysis

Data analysis is another challenging step. First it is crucial to examine the reproducibility of the data. As discussed above, the variability observed in the results is due to variations in the biological material, in sample handling during RNA, cDNA and cRNA preparation, and in hybridization performance. To eliminate the latter source of variability, intensities can be normalized by linear transformation (y=a*x+b). Several approaches can be envisaged to set parameters a and b. One is to correct the original values so that the background and the mean intensity of fluorescence are the same for all chips; another is based on the intensity of reference probes present on all chips; one can also aim to set the central values of the intensity distribution in a predetermined range of values, for example such that the 10th and 90th percentiles are always equal to identical values. After normalization, statistical analysis of the data can be performed. Such analyses should determine the level of significance of the overall variation of expression levels measured between two conditions. A method based on principal component analysis has been described as a practical approach to determine which genes show statistically significant altered expression (Hilsenbeck et al. 1999) based on the overall variability of expression intensities between two samples. Very often, intensities obtained for weakly expressed genes are less reliable (and therefore more variable) than values obtained for highly expressed genes. In such cases, it is necessary to determine the limit of significance for classes of data along the first component axis rather than for the whole data collection (Fig. 5.2). Thus even small changes of expression levels of the highly expressed genes can be statistically significant, whereas only strong variations of expression level will be significant for weakly expressed genes.

However, comparing gene expression under two conditions is not satisfactory for most research projects. Indeed, numerous studies have used hierarchical clustering and color-coding of expression ratios to visualize expression of a large number of genes under many conditions (for example, Chu et al. 1998; Eisen et al.. 1998; Wen et al. 1999; Alon et al. 1999). The resulting images provide an efficient way for the human brain to grasp expression patterns characterizing groups of genes and/or conditions. More sophisticated algorithms have also been developed to achieve nonhierarchical clustering of data in order to avoid biases dependent on the order of data input, and artefacts due to the hierarchical concept. These methods (k-mean clustering; Tavazoie et al. 1999; Self-Organizing Maps (SOM); Golub et al. 1999; Tamayo et al. 1999; Törönen et al. 1999) appear to be efficient at defining functional groups of genes. The methods described above are all based on defining similarity measurements between expression patterns, with no prior knowledge of the functional classes. Most recently, a supervised learning technique [Support Vector Machines (SVM); Brown et al. 2000] has been described for clustering expression data. A training set of data is first used to teach the SVM to discriminate between members of different classes. The trained SVM is then able to classify other genes among these classes.

A hierarchical clustering program (Eisen et al. 1998) can be downloaded at no cost to public laboratories from http://rana.stanford.edu/software. Links to other sites providing noncommercial software analysis tools can be found at http://industry.ebi.ac.uk/~alan/microarray. Numerous companies are also developing data analysis software at a rapid rate. For a comprehensive and up-

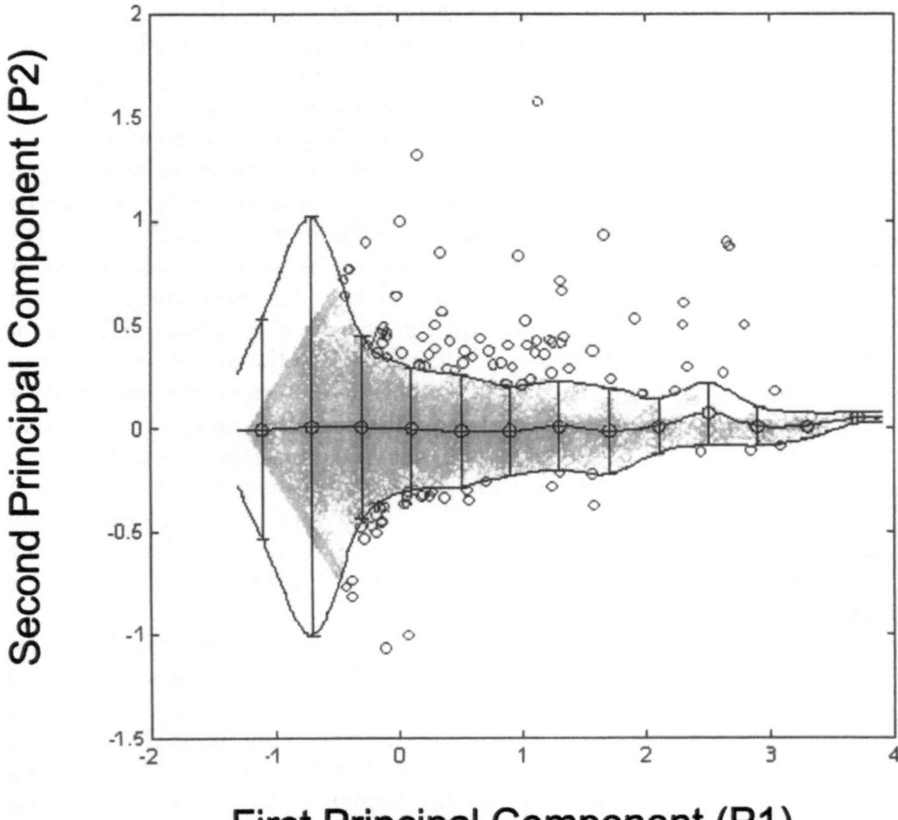

Fig. 5.2. Comparative gene expression in two conditions using principal component analysis. Expression intensities for each gene were averaged after normalization of signals obtained from four DNA chips hybridized with either control or treated cRNA. To determine which transcripts were significantly affected by treatment, statistical analysis was performed using principal component analysis as described in Hilsenbeck et al. (1999). First principal component (*P1*) represents expression intensities. Second principal component (*P2*) represents difference in expression between the two studied conditions. Data have been log-transformed. Contiguous classes of data along component P1 axis were used to calculate limit values, which were joined by interpolation to construct the 99.5% prediction region for component P2. Genes showing significant differences in expression are represented by *dark gray dots* outside the upper (increased expression) and lower (decreased expression) limit curves

dated list of links to their web sites, consult the Gene Expression Technology Group web site from Wake Forest School of Medicine (http://www.wfubmc.edu/physpharm/genetech/genetechlinks.html). Several demo versions of commercial clustering programs can be downloaded (GenExplore from Applied Maths: http://www.applied-maths.com/ge/ge.html; GeneSpring from Silicon Genetics: http://www.sigenetics.com/products/genespring/index.html; Spotfire Pro from Spotfire: http://www.spotfire.com/products/default.asp).

3.3 Interpretation of Results

Finally, once an overall view of the regulated genes has been obtained, one wishes to understand the mechanism of action of the gene products involved. A prerequisite is to access as much information as possible on the modulated genes. To this aim, one needs an efficient way to consult databases containing genes definition, protein characteristics, function, etc. Yet, since several hundreds of public databases are presently available on the Internet, one faces the problem of integrating all the available data. This step is extremely time consuming.

Several databases with integrated information on genes, proteins and pathways have been recently developed and will greatly facilitate the interpretation of gene expression studies:

1. The BioKnowledge Library is a growing collection of databases developed by Proteome Inc. which provides all existing information on proteins of several organisms. Available information has been gathered from the literature by human curators, and synthetic knowledge on each protein is presented in a single Web page format, organized by topics such as function, chemical and physical properties, sequence, bibliographic references, etc. Hypertext links are used to relate information in each field to BioKnowledge pages concerning related proteins as well as to public databases. Proteome's *Saccharomyces cerevisiae* and *Caenorhabditis elegans* databases (Costanzo et al. 2000) are freely accessible to academic laboratories at http://www.proteome.com/databases/index.html. Three other databases are accessible only upon subscription. One covers *Candida albicans* proteins, the other concerns G-protein coupled receptors and associated proteins, and the last one describes more than 17,000 human proteins and predicted proteins, also referring to mouse and rat proteins.
2. Another valuable integrated database for human genes is GeneCards (Rebhan et al. 1998). For more than 9,000 human genes, an 'identity card' is available displaying comprehensive information, including cellular functions of the gene products and their medical implications, as well as links to entries in other databases. Information displayed on the cards has

been compiled computationally from several major public databases. GeneCards is available at http://bioinformatics.weizmann.ac.il/cards.
3. For *Saccharomyces cerevisiae*, summaries of published information for each known gene and its products are progressively introduced in the *Saccharomyces* Genome Database (Ball et al. 2000), available at http:// genome-www.stanford.edu/saccharomyces.

In addition, several databases oriented towards metabolic pathways are available. The KEGG project (Kanehisa and Goto 2000) relies on a multi-species conceptualization of pathways, whereas the WIT (Overbeek et al. 2000) and DoubleTwist (previously Pangea; Karp et al. 2000) databases define species-specific pathways. The KEGG project includes most known metabolic pathways as well as membrane transport, signal transduction and cell cycle, concerning genes from a wide variety of sequenced genomes. Access is freely available at http://www.genome.ad.jp/kegg/kegg2.html. WIT bears on a wider range of pathways, including DNA and protein synthesis and modification, signal transduction, transmembrane transport, etc. from over 40 micro-organisms and *Caenorhabditis elegans*. The database is freely available at http:/ /wit.mcs.anl.gov/wit2. The DoubleTwist metabolic databases consist of two products mainly focused on metabolic pathways of microorganisms. EcoCyc concerns *E. coli* small-molecule metabolism and membrane transport systems. A software tool allows visualization of gene expression data on the whole metabolic network of the organism. MetaCyc contains the EcoCyc pathways as well as pathways from a variety of other microorganisms including *Saccharomyces cerevisiae* and a few plant and human pathways. The *E. coli* database has been created through manual curation, whereas MetaCyc was created computationally and extended manually. An account (free to academics) is required to access the EcoCyc/MetaCyc Web server. Information is available at http://ecocyc.pangeasystems. com.

4 Conclusion

Large-scale expression measurement here exemplified by the Affymetrix technology is providing new perspectives in biology. Global expression analysis is still in its infancy, but now many reports illustrate diverse applications that highlight the power of this tool. Recent expression profiling examples include global monitoring of gene expression associated with cancer – lung (Petersen et al. 2000), prostate (Carlisle et al. 2000) or colon cancer (Alon et al. 1999), with pathological processes (rheumatoid arthritis, Heller et al. 1997); infection (HIV, Geiss et al. 2000); normal physiological process (central nervous system development, Wen et al. 1999); cytokine-mediated activation of T and B lymphocytes (Alizadeh et al. 1998); implantation (Yoshioda et al. 2000); p53-regulated genes (Zhao et al. 2000); etc. In drug discovery, expression profiling technology also dramatically influences pharmaceutical strategies in drug development programs, either for drug validation or for identification of secondary drug targets. Indeed, this technology is revolutionizing toxicological analyses as such global analyses provide a rapid definition of the signature of a molecule by the identification of all the modulated targets. It also allows a rapid detection of undesirable secondary effects of the new drugs under study (Marton et al. 1998).

The few examples illustrate the spreading of the technology through all fields of biology. Oligonucleotide chip technology is a promising tool that will undoubtedly help to understand biological processes where difference in gene expression is of significance and will offer the opportunity to develop new efficient and original therapies.

The example described here has been published in Derocq et al. (2000).

References

Alizadeh A, Eisen M, Botstein D (1998) Probing lymphocyte biology by genomic-scale gene expression analysis. J Clin Immunol 18(6):373–379

Alon U, Barkai N, Notterman DA (1999) Broad patterns of gene expression revealed by clustering analysis of tumor and normal colon tissues probed by oligonucleotide array. Proc Natl Acad Sci USA 96(12):6745–6750

Ball CA, Dolinski K, Dwight SS, Harris MA, Issel-Tarver L, Kasarskis A, Scafe CR, Sherlock G, Binkley G, Jin H, Kaloper M, Orr SD, Schroeder M, Weng S, Zhu Y, Botstein D, Cherry JM (2000) Integrating functional genomic information into the *Saccharomyces* genome database. Nucleic Acids Res 28(1):77–80

Bouaboula M, Rinaldi-Carmona M, Carayon P, Carillon C, Delpech B, Shire D, Le Fur G, Casellas P (1993) Cannabinoid-receptor expression in human leukocytes. Eur J Biochem 214:173–180

Brown MP, Grundy WN, Lin D, Cristianini N, Sugnet CW, Furey TS, Ares M Jr, Haussler D (2000) Knowledge-based analysis of microarray gene expression data by using support vector machines. Proc Natl Acad Sci USA 97(1):262–7

Carlisle A, Prabhu V, Elkahloun A (2000) Development of a prostate cDNA microarray and statistical gene expression analysis package. Mol Carcin 28:12–22

Caulfield MP, Brown DA (1992) Cannabinoid receptor agonists inhibit Ca current in NG108-15 neuroblastoma cells via a pertussis toxin-sensitive mechanism. Br J Pharmacol 106(2):231–232

Childers SR, Pacheco MA, Bennett BA, Edwards TA, Hampson RE, Mu J, Deadwyler SA (1993) Cannabinoid receptors: G-protein-mediated signal transduction mechanisms. Biochem Soc Symp 59:27–50

Chomzcynski P, Sacchi N (1987) Single-step method of RNA isolation by acid-guandium-thiocyanate-phenol-chloroform extraction. Anal Biochem 162:156–159

Chu S, DeRisi J, Eisen M, Mulholland J, Botstein D, Brown PO, Herskowitz I (1998) The transcriptional program of sporulation in budding yeast. Science 282:699–705

Cohen P, Bouaboula M, Bellis M, Baron V, Jbilo O, Poinot-Chazel C, Galiegue S, Hadibi EH, Casellas P (2000) Monitoring cellular responses to *Listeria monocytogenes* with oligonucleotide arrays. J Biol Chem 275:11181–11190

Costanzo MC, Hogan JD, Cusick ME, Davis BP, Fancher AM, Hodges PE, Kondu P, Lengieza C, Lew-Smith JE, Lingner C, Roberg-Perez KJ, Tillberg M, Brooks JE, Garrels JI (2000) The yeast proteome database (YPD) and *Caenorhabditis elegans* proteome database (WormPD): comprehensive resources for the organization and comparison of model organism protein information. Nucleic Acids Res 28:73–76

Cross AK, Richardson V, Ali SA, Palmer I, Taub DD, Rees RC (1997) Migration responses of human monocytic cell lines to alpha- and beta-chemokines. Cytokine 9:521–528

Derocq JM, Jbilo O, Bouaboula M, Segui M, Clere C, Casellas P (2000) Genomic and functional changes induced by the activation of the peripheral cannabinoid receptor CB2 in the promyelocytic cells HL-60. Possible involvement of the CB2 receptor in cell differentiation. J Biol Chem 275:15621–15628.

Devane WA, Dyzarz III FA, Johnson MR, Melvins LS, Howlett AC (1988) Determination and characterization of a cannabinoid receptor in rat brain. Mol Pharmacol 34;605–613

Eisen MB, Spellman PT, Brown PO, Botstein D (1998) Cluster analysis and display of genome-wide expression patterns. Proc Natl Acad Sci USA 95:14863–14868

Galiègue S, Mary S, Marchand J, Dussossoy D, Carrière D, Carayon P, Bouaboula M, Shire D, Le Fur G, Casellas P (1995) Expression of central and peripheral cannabinoid receptors in human immune tissues and leukocyte subpopulations. Eur J Biochem 232:54–61

Geiss GK, Bumgarner RE, An MC (2000) Large-scale monitoring of host cell gene expression during HIV-1 infection using cDNA microarrays. Virology 266:8–16

Gérard CM, Mollereau C, Vassart G, Parmentier M (1991) Molecular cloning of a human cannabinoid receptor which is also expressed in testis. Biochem J 279:129–134

Golub TR, Slonim DK, Tamayo P, Huard C, Gaasenbeek M, Mesirov JP, Coller H, Loh ML, Downing JR, Caligiuri MA, Bloomfield CD, Lander ES (1999) Molecular classification of cancer: class discovery and class prediction by gene expression monitoring. Science 286:531–537

Heller R, Schena M, Chai A, Shalon D, Bedilion T, Gilmore J, Wooley D, Davis R (1997) Discovery and analysis of inflammatory disease-related genes using cDNA microarrays. Proc Natl Acad Sci USA 94:2150–2155

Hilsenbeck SG, Friedrichs WE, Schiff R, O'Connell P, Hansen RK, Osborne CK, Fuqua SAW (1999) Statistical analysis of array expression data as applied to the problem of tamoxifen resistance. J Natl Cancer Inst 91:453–459

Howlett AC (1995) Pharmacology of cannabinoid receptors. Annu Rev Pharmacol Toxicol 35:607–634

Kaminski NE, Abood ME, Kessler FK, Martin BR, Schatz AR (1992) Identification of a functionally relevant cannabinoid receptor on mouse spleen cells that is involved in cannabinoid-mediated immune modulation. Mol Pharmacol 42:736–742

Kanehisa M, Goto S (2000) KEGG: Kyoto encyclopedia of genes and genomes. Nucleic Acids Res 28:27–30

Karp P, Riley M, Saier M, Paulsen IT, Paley SM, Pellegrini-Toole A (2000) The EcoCyc and MetaCyc databases. Nucleic Acids Res 28:56–59

Marton M, et al. (1998) Drug target validation and identification of secondary drug target effects using DNA microarrays. Nat Med 4(11):1293–1301

Matsuda LA, Lolait SJ, Brownstein MJ, Young AC, Bonner TI (1990) Structure of a cannabinoid receptor and functional expression of the cloned cDNA. Nature 346:561–564

McGall G, Labadie J, Brock P (1996) Light-directed synthesis of high-density oligonucleotide arrays using semiconductor photoresistance. Proc Natl Acad Sci USA 93:13555–13560

Munro S, Thomas KL, Abu-Shaar M (1993) Molecular characterization of a peripheral receptor for cannabinoids. Nature 365:61–65

Overbeek R, Larsen N, Pusch GD, D'Souza M, Selkov E Jr, Kyrpides N, Fonstein M, Maltsev N, Selkov E (2000) WIT: integrated system for high-throughput genome sequence analysis and metabolic reconstruction. Nucleic Acids Res 28:123–125

Petersen S, Heckert C, Rudolf J, Schlüns K, Tchernitsa O, Schäfer R, Dietel M, Petersen I (2000) Gene expression profiling of advanced lung cancer. Int J Cancer 86:512–517

Rebhan M, Chalifa-Caspi V, Prilusky J, Lancet D (1998) GeneCards: a novel functional genomics compendium with automated data mining and query reformulation support. Bioinformatics 14:658–664

Rinaldi-Carmona M, Barth F, Millan J, Derocq J-M, Casellas P, Congy C, Oustric D, Sarran M, Bouaboula M, Calandra B, Portier M, Shire D, Brelière J-C, Le Fur G (1998) SR 144528, the first potent and selective antagonist of the CB2 cannabinoid receptor. J Pharmacol Exp Ther 284:644–650

Schatz AR, Kessler FK, Kaminski NE (1992) Inhibition of adenylate cyclase by delta 9-tetrahydrocannabinol in mouse spleen cells: a potential mechanism for cannabinoid-mediated immunosuppression. Life Sci 51(6)25–30

Tamayo P, Slonim D, Mesirov J, Zhu Q, Kitareewan S, Dmitrovsky E, Lander ES, Golub TR (1999) Interpreting patterns of gene expression with self-organizing maps: methods and application to

hematopoietic differentiation. Proc Natl Acad Sci USA 96:2907–2912
Tavazoie S, Hughes JD, Campbell MJ, Cho RJ, Church GM (1999) Systematic determination of genetic network architecture. Nat Genet 22:281–285
Törönen P, Kolehmainen M, Wong G, Castrén E (1999) Analysis of gene expression data using self-organizing maps. FEBS Lett 451:142–146
Wen X, Fuhrman S, Michaels GS (1999) Large-scale temporal gene expression mapping of central nervous system development. Proc Natl Acad Sci USA 95(1):334–339
Yoshioda K, Matsuda F, Takahura K, Noda Y, Imakawa K, Sakai S (2000) Determination of genes involved in the process of implantation: application of GeneChip to scan 6,500 genes. Biochem Biophys Res Comm 272:531–538
Zhao R, Gish K, Murphy M, Yin Y, Notterman D, Hoffman W, Tom E, Mack D, Levine A (2000) Analysis of p53-regulated gene expression patterns using oligonucleotide arrays. Genes Dev 14:981–993

CHAPTER 6

Gene Expression Data Mining and Analysis

Alvis Brazma[1], Alan Robinson[1] and Jaak Vilo[1]

[1] European Bioinformatics Institute, Hinxton, Cambridge CB10 1SD, UK

1 Introduction

Using microarrays, we can build a table of gene expression profiles that characterize the dynamic functioning of each gene in the genome by measuring gene transcription levels at different developmental stages, in different tissues, and under various conditions. Rows in this table represent genes, columns represent samples, such as different tissues, developmental stages and treatments, and each position in the table contains values characterizing the expression level of that particular gene in a particular sample. We call this table a *gene expression matrix.*

Building up a database of gene expression matrices from different microarray experiments may help us to understand gene regulation, metabolic and signaling pathways, the genetic mechanisms of disease, and the cellular response to toxins and drug treatments (e.g., see Sander 2000; Young 2000). For instance, it can help in:

- prediction of gene function for uncharacterized genes based on the similarity of their expression profiles to those of known genes;
- identifying genes important in specific cellular processes, diseases, or in cell differentiation;
- finding how cells respond to various compounds and then classifying these for prediction of responses by new compounds (this approach is especially important for toxicology studies);
- learning about gene regulation by finding and studying groups of co-regulated genes.

More generally, microarrays are changing the very way in which a complex biological system like a cell can be viewed. By taking snapshots of the gene expression levels of virtually all known genes at a given moment, microarrays are transforming a complex biological system like a cell from a black box to a transparent box. The amount of information that a well-staged microarray experiment can reveal is potentially huge, but it is not a trivial task to extract information and ultimately knowledge from these data. The amounts of data are too large to study all the data points or genes individually, and the traditional "one gene at a time" approaches are no longer applicable. Instead, we must have automated techniques to pre-filter the data and find the most interesting genes that can then be studied in more detail. The new opportunities created by microarray data are tremendous. The traditional hypothesis-driven research in biology is no longer the only possibility. Data mining of gene expression data matrices can suggest potential new targets automatically, which may then be explored by more traditional methods for verification.

There are two relatively distinct stages of microarray data analysis: (1) raw data analysis – transforming the scanned microarray images into quantitative data about gene expression, i.e. obtaining a gene expression data matrix from the raw data; and (2) gene expression matrix analysis – analyzing the obtained quantitative data in order to understand the underlying biological processes.

In the next section we will start with a discussion of the approaches for the raw data analysis stage, while the rest of the paper will be devoted to methods of gene expression matrix analysis.

2
From Raw Data to Gene Expression Matrix

Microarrays measure the relative or absolute mRNA abundance indirectly by measuring the intensity of the fluorescence of the labeled mRNA bound to spots on the array (e.g. see Duggan et al. 1999). For each fluorescent dye the intensities at an appropriate wavelength are measured on a separate channel, and the raw data produced by microarrays are in fact monochrome images for each channel (Fig. 6.1). Transforming these images into the gene expression matrix is not a trivial task. The spots corresponding to genes on the microarray have to be identified, their boundaries determined, the fluorescence intensity from each spot measured and compared to the background intensity and then to the intensities for other channels. The software for this initial image processing is often provided with the image scanner, since it will depend on the particular hardware. For some image analysis packages, laborious manual adjustment of the grid for spots is necessary to ensure the quality of analysis. The main features desirable for image analysis software are automatic image overlaying, gridding and spot finding, and signal quantization.

One of the most popular software packages for image analysis from spotted arrays is ScanAlyze (available from http://rana.stanford.edu/software. Developed by Michael Eisen while at Stanford University, now at UC Berkeley), this package includes semi-automatic definition of grids and complex pixel and spot analyses. The derived intensities may be output as tab-delimited text files for transfer to any database or analysis application. The package is available as a binary for Windows, but source code is also available for porting to other platforms.

Most of the microarray image analysis tools used currently are largely in agreement about how to quantify "good" spots. The problem, however, is to identify defective spots, such as doughnut-shaped spots, or imperfections on the surface of the array automatically and reliably. Even if such imperfections may be identified either automatically or manually, it is not always obvious how to quantify them. However, completely discarding information from such imperfect spots is considered as wasteful currently, so research continues in this area.

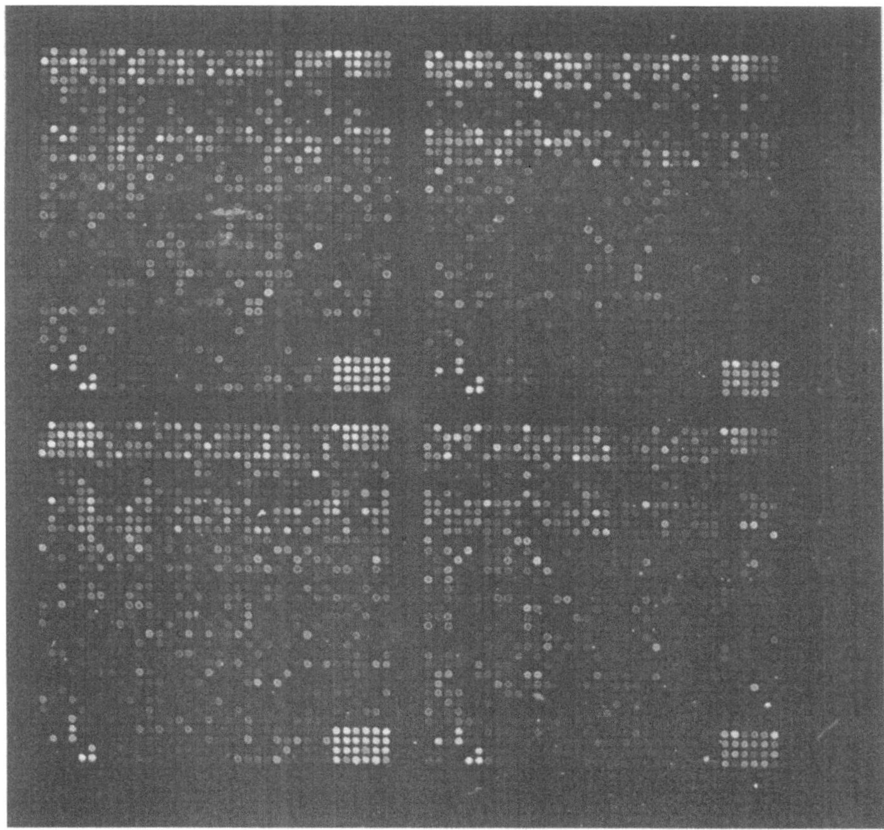

Fig. 6.1. Sample image from scanning a hybridized rat microarray containing over 5,000 genes. Each spot features a pool of identical single-stranded DNA molecules representing a single gene. Brightness of spot is proportional to amount of fluorescent mRNA hybridized to DNA of spot. Automated image analysis software should identify these fluorescence spots, determine their boundaries, and fluorescence intensity from each spot should be measured and compared to background fluorescence. Moreover, the image should be compared to a similar image obtained from control measurements and ratio of background subtracted intensities calculated. In this way images are transformed into the gene expression matrix, which can be analyzed further by numerical methods. (Image was kindly provided by Tom Freeman, Sanger Centre, Cambridge, UK)

A survey of image analysis software can be found at http://cmpteam4.unil.ch/biocomputing/array/software/MicroArray_Software.html. We will not discuss concrete packages in detail here, because this is a field that is developing extremely rapidly.

In any experiment it is important to know not only the value of the measurement, but also a standard error or some other reliability indicator for each data-point. For most microarray technology platforms only the ratio of background-subtracted signals between the given sample and the control sample is meaningful. However, if the control spot intensity is low, then the ratio of these numbers may be high and thus appear significant, although this measurement may not be reliable because of the difficulty of determining intensities accurately at low levels. The quality of the spot may be assessed using many factors, such as the uniformity of the individual pixel intensities across the spot, or the shape of the spot. Unfortunately, there is no standard way of assessing the reliability of spot measurements or measurement ratios. If experiments have been done as replicates, then replicates for each spot may be used to calculate standard errors for consolidated spot measurements and hence improve quality assessment. However, little has been published yet on how to assess the reliability of gene expression measurements obtained by combining information from the spot image in each channel and the replicate images, yet their importance has been noted by many authors. Recent studies of replicate experiments (Lee et al. 2000) demonstrate the possible unreliability of single hybridization measurements.

Schuchhardt et al. (2000) discuss the major sources of fluctuations to be expected in microarray experiments and compare four different normalization procedures. The question of how to take these reliability indicators and error measurements into account during analysis remains one of the many challenges in the analysis of microarray gene expression data.

Another difficulty when studying a gene expression matrix comes from the necessity to link the DNA printed on each spot with its respective gene. This is not always possible, since the DNA used is derived from EST sequences typically, and linking an EST to its respective full-length gene may be nontrivial. Usually this is achieved through EST clustering. Furthermore, the same gene may be represented by several spots on the array, either by exactly the same sequence or by (non-homologous) sequences from different parts of the gene. If measurements from these different spots differ, then it must be considered how they should be consolidated to a single expression level for that gene.

Currently, microarray-based gene expression measurements are relative by nature: essentially, we can compare meaningfully only the expression level of the same gene in different samples or different genes in the same sample. Moreover, an appropriate normalization should be applied to enable any comparison of data. Unfortunately, the different normalization techniques used currently (e.g. comparison to foreign “spiked” genes, or comparison to housekeeping genes which are assumed to have stable expression in all tissues and under all conditions, if they exist) can produce radically different results.

It is typically assumed that an abundance ratio of greater than 1.5 or 2 is indicative of a significant change in gene expression, but such estimates are very crude since the reliability of ratios depends on the absolute intensity values. The intensity of the spot can be affected by many factors, including the specificity of the sequence and cross-hybridization of other homologous sequences to the spot. Secondly, given the large number of measurements, we should expect a proportion of measurements to be incorrect on statistical grounds. In practice this proportion could mean that tens or even hundreds of genes out of tens of thousands are showing significant differential expression purely on the basis of statistical phenomena (for instance see Claverie 1999). This should be kept in mind when analyzing gene expression matrices.

Finally, it should be noted that microarray image analysis methods are still regarded as one of the major bottlenecks limiting the reliability of the technology. This process will become more reliable with the next generation of microarrays, where all the spots will be printed (or synthesized) in several replicates rather than singly. If there are enough replicates for each spot on the array, then the measurements from imperfect spots can be discarded. Secondly, replicates will allow us to assign error bars to measurements for genes by calculating the standard deviation from the individual measurements of each reliable spot.

With more and more genomes of different species being sequenced and genes identified, it will become feasible to build "gene centric" instead of "spot centric" expression databases. Each gene in the genome will be represented by one or more oligonucleotides that are characteristic for that particular gene. Where there are homologous genes, appropriate sequences should be used to distinguish between them. The information from several spots representing the same gene will have to be integrated in the analysis. Note that Affymetrix oligo chips use a similar approach already.

Having processed the raw image data into a gene expression matrix, we need to ascertain new information concerning the underlying biological processes.

3 Gene Expression Matrix Analysis

Gene expression matrices can be used in many ways. We can distinguish between supervised and unsupervised data analysis, as well as between hypotheses-driven analysis and data mining, with the aim of generating new hypotheses. The supervised approach assumes that for some (or all) profiles we have additional information, such as functional classes for the genes, or diseased/normal states attributed to the samples. We can view this additional information as labels attached to the rows or columns. Having this information, a typical task is to build a classifier that is able to predict the labels from an expression profile of an unclassified gene or sample. A typical example of an unsupervised data analysis is clustering expression profiles to find groups of co-regulated genes or related samples.

An example of a hypothesis-driven analysis is to pick a potentially interesting gene (e.g. a gene that is known to be related to a certain disease), then find a group of genes with similar or anti-correlated expression profiles by expression data comparisons. This group can then be studied further. A simple example of the data mining approach is looking for genes differentially expressed in various samples.

Another way of looking at the possibilities of how a gene expression matrix can be analyzed is based on either a gene-centric or sample-centric view, or a combined view. Concretely: a gene-centric view compares expression profiles of genes by comparing rows in the expression matrix; a sample-centric view compares expression profiles of samples by comparing columns in the expression matrix. Additionally, both methods can be combined (provided that the data normalization method used allows these comparisons).

When comparing rows or columns, we can look for either similarities or differences. If we find that two rows are similar, we can hypothesize that the two respective genes are co-regulated and hence possibly functionally related. By comparing samples, we can find which genes are differentially expressed and, for instance, study the effects of various compounds.

Multivariate techniques, such as principal component analysis (PCA), can also be used for the analysis. Usually the aim is to reduce the dimensionality of the data space to the few most important dimensions using a linear combination of the original dimensions. This allows the subsequent visualization and analysis in a lower-dimensional space of the original expression data.

As many expression studies capture the changes in cells during some biological process, these measurements are often time courses measuring gene expression at specific time points. For time-series analysis, the number of analysis methods available is rich as these type of data have been studied extensively in other fields, as for example in stock-market analysis and signal processing. Fourier analysis is among these methods.

3.1 Expression Profile Similarity Measures

To be able to perform comparisons of expression profiles, we need a method to measure the similarity (or distance) between the objects we are comparing. We can regard the expression profiles (either rows or columns in the matrix) as points in n-dimensional space or as n-dimensional vectors, where n is either the number of samples for gene comparison or the number of genes for sample comparison. The natural, so-called Euclidean distance between these points in the *n*-dimensional space may be the most obvious choice, but not necessarily the best one. Let $A=(a_1,\dots,a_n)$ and $B=(b_1,\dots,b_n)$ be two *n*-dimensional points. Then the Euclidean distance is defined as:

$$D_{eucl}(A,B)=\sqrt{\sum_{i=1}^{n}(a_i-b_i)^2}$$

Intuitively appealing is to use the correlation coefficient calculated by treating the two *n*-dimensional vectors as a series of random variables. In fact, this distance is related to the angle between the two *n*-dimensional vectors and its simplest version can be calculated as follows:

$$D_{corr}(A,B)=1-\frac{\sum_{i=1}^{n}(a_i-\overline{a}_i)(b_i-\overline{b}_i)}{\sqrt{\sum_{i=1}^{n}(a_i-\overline{a}_i)^2\sum_{i=1}^{n}(b_i-\overline{b}_i)^2}},$$

where *ai* and *bi* are mean values of a_i and b_i for $i=1,\dots,n$, respectively. Euclidean and correlation distance measures are related if we normalize the length of the vectors to 1. If we assume that the mean values equal to 0, then the *chord distance* can be expressed as follows:

$$D_{chord}(A,B)=\sqrt{2\left(1-\frac{\sum_{i=1}^{n}a_ib_i}{\sqrt{\sum_{i=1}^{n}a_i^2\sum_{i=1}^{n}b_i^2}}\right)}$$

This allows us to use a correlation distance even for the cases where Euclidean properties are important, for instance for *K*-means clustering algorithm (Legendre and Legendre 1998).

Some other distance measures, including rank correlation coefficient and mutual information-based measures, are proposed in D'haesleer et al. (1998). Currently, to the best of our knowledge, there is no theory on how to choose the most appropriate distance measure. Possibly one single "right" distance measure for expression profile space does not exist, and the choice will depend on the questions that we ask and the properties of the underlying data. Standard sets of known co-regulated genes in various organisms and gene regulatory network modelling may help in finding theoretically substantiated similarity measures and establishing guidelines for the best choice of method.

After having chosen the similarity measure for the expression profile space, we can study the expression matrix in either a supervised or an unsupervised manner.

3.2 Clustering

Clustering is the most popular method currently used in the first step of gene expression matrix analysis. The goal of clustering is to group together objects (i.e., genes or samples) with similar properties. This can be viewed also as a reduction of the dimensionality of the system, allowing the reduction of tens of thousands of genes to a few groups each containing similarly behaving genes. Clustering is not a new technique, many algorithms have been developed for it (for review see Jain et al. 1999), and many of these algorithms have been applied to analyze expression data. The hierarchical (Eisen et al. 1998; Alon et al. 1999) and *K*-means clustering algorithms (Hartigan 1975; Tavazoie et al. 1999; Vilo et al. 2000), as well as self-organizing maps (Tamayo et al. 1999; Toronen et al. 1999), have all been used for clustering expression profiles. Even a simple clustering algorithm based on binning (i.e. discretizing the expression profile space and clustering together the profiles that map into the same bin) has been shown to be useful for gene clustering (assuming that there are not too many experimental conditions) and subsequent discovering of transcription factor binding sites (Brazma et al. 1998). More recently, new algorithms have been developed specifically for gene expression profile clustering (for instance Ben-Dor et al. 1999; Sharan and Shamir 2000) based on finding approximate cliques in graphs.

Hierarchical clustering works by iteratively joining together the two closest clusters, starting from singleton clusters (Eisen et al. 1998) or iteratively partitioning clusters starting with the complete set (Alon et al. 1999). After each joining of two clusters, the distances between all the other clusters and the new joined cluster are re-calculated (see Fig. 6.2). The complete linkage, average linkage and single linkage methods use maximum, average and minimum distances between the members of two clusters respectively. Note that to obtain a particular partitioning into clusters, the threshold distance should be chosen by independent means (typically by the user using visual inspection).

The *K*-means clustering algorithm typically uses the Euclidean properties of the vector space. The desired number of clusters *K* has to be chosen a priori. After the initial partitioning of the vector space into *K* parts, the algorithm calculates the centre point of each subspace and adjusts the partition so that each vector is assigned to the cluster to whose center it is closest (see Fig. 6.3). This process is repeated until either the partitioning stabilizes or a given number of iterations is exceeded. The approaches for the initial selection of the first set of *K* cluster centers can vary. One possibility is to use a random selection of vectors as the original cluster centers, and possibly repeat this several times to sample different starting points. Alternatively, centres can be chosen deterministically by first defining a centroid vector and then selecting the $K-1$ most distant vectors. Yet another possibility is to choose the *K* centres after a hierarchical clustering of a small sample of vectors (Legendre and Legendre 1998).

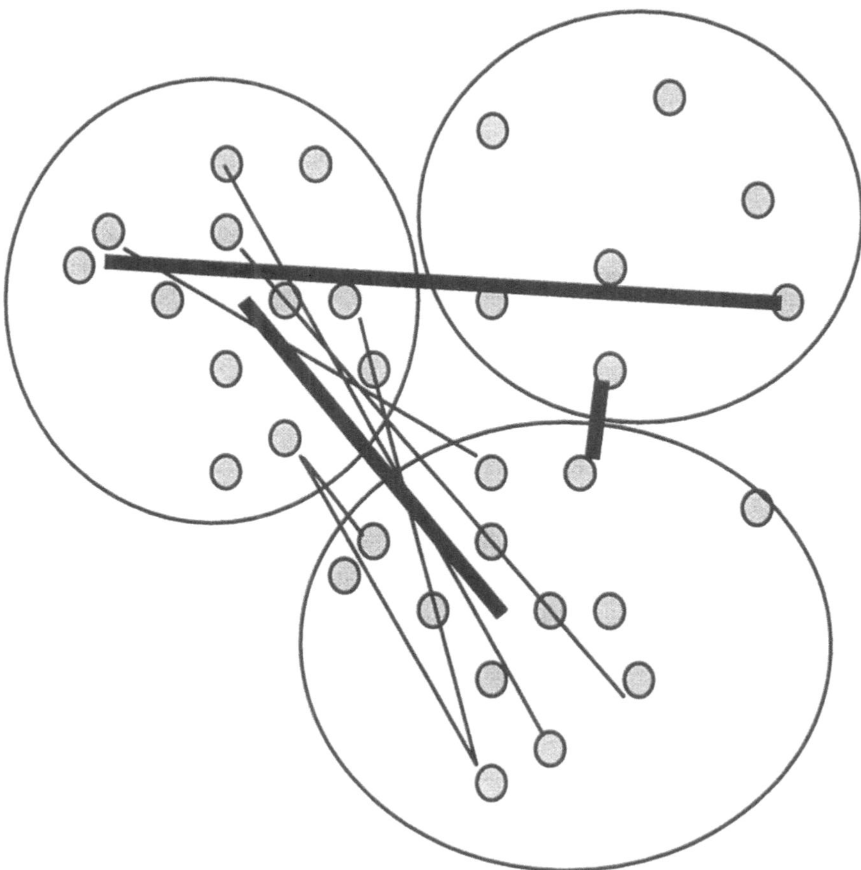

Fig. 6.2. Hierarchical clustering by iteratively joining the two clusters that are closest to each other, starting from singleton clusters. After each joining of two clusters, distances between all other clusters and new joined cluster are recalculated. Complete linkage, average linkage and single-linkage methods use maximum, average and minimum distances between members of two clusters respectively, as shown in figure. In this way a tree-like structure of clusters covering the object space is created. To obtain any particular clustering system the user has to choose a cut-off distance, when joining is stopped

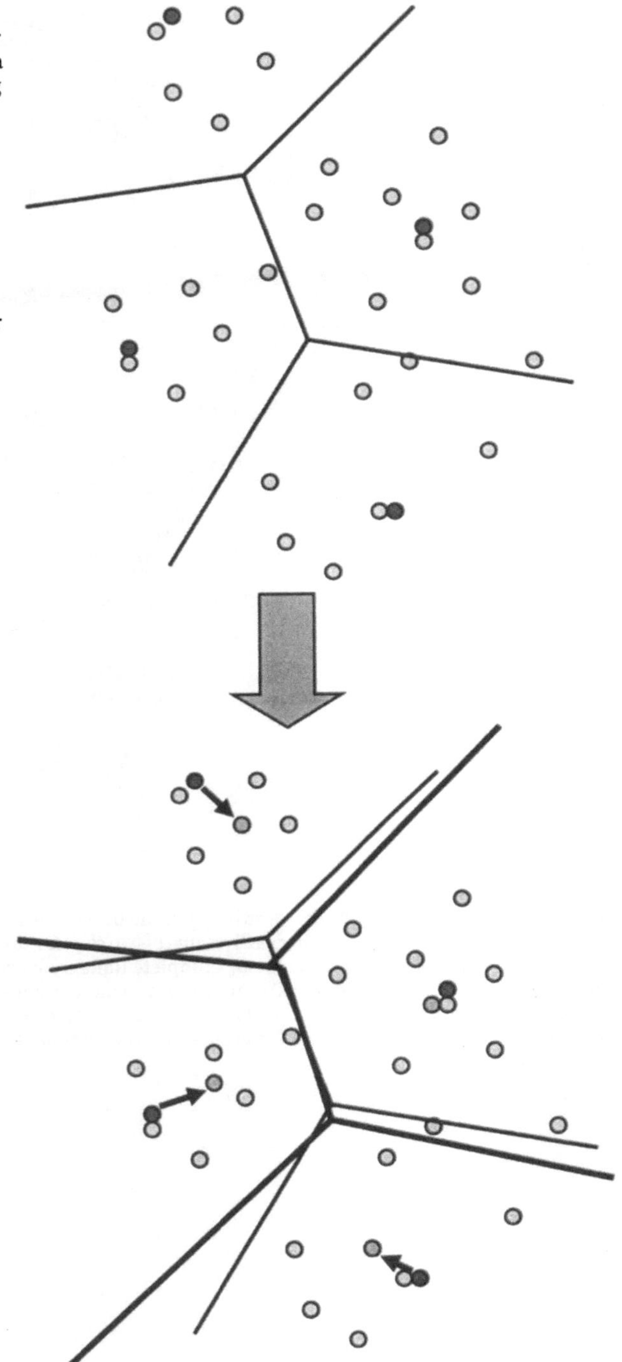

Fig. 6.3. For K-means clustering, the desired number of clusters K has to be chosen a priori. After initial partitioning of vector space into K parts (which can be done in various ways, for instance by random partitioning), the algorithm calculates the centre point of each subspace and adjusts the partition so that each vector is assigned to the cluster to whose centre it is closest. This process is repeated until either the partitioning stabilizes or a given number of iterations is exceeded

A novel method of "gene shaving" (Tibshirani et al. 1999) provides a way of combining the two-way clustering with a combination of supervised and unsupervised analysis. This clustering method relies on an iterative heuristic algorithm which tries to identify subsets of genes with coherent expression patterns and a large variation across conditions. The technique can be used in unsupervised, partially supervised or fully supervised analysis by utilizing the known properties of the genes or samples. The algorithm works iteratively by alternating the PCA analysis and excluding the least variable objects in each step. The "best cluster" and the number of clusters are identified using a novel statistical criteria called the "Gap statistic".

A novel graph theory-based clustering algorithm, called CLICK, designed specifically for gene expression profile clustering has been proposed in (Ben Dor et al. 1998). This algorithm does not require any prior assumptions on the structure or the number of the clusters. The algorithm has been tested on a variety of biological data sets and has achieved a good performance.

An alternative to selecting a particular clustering method is to study a number of different clusterings in parallel. A *K*-means clustering algorithm has been used in this way in (Vilo et al. 2000). By repeating clustering for many different *K* values as well as varying the initializations for each *K*, about 52,000 different (but overlapping) clusters together with a formal "goodness" measure for each one were created for further studies (see Sect. 6).

Questions about how many clusters there are in the data and how good one or another clustering technique is are relevant; however, they lack simple answers. There are some proposed measures and techniques to tackle these questions, but none of these is universal. A possible approach for estimating the number of clusters would be the methods using the Minimum Description Length principle (MDL; Li and Vitanyi 1993).

The "goodness" of a cluster depends on how close its elements are to each other, and how far they are from the next closest cluster. One such measure has been proposed by Rousseeuw (1987) based on the notion of a *silhouette plot* and an average silhouette value of a cluster defined as follows. For each two objects i and j, we denote by $d(i;j)$ the distance between i and j. For a set A, we denote by $|A|$ the number of elements in A. For each object i we denote by A the cluster to which it belongs and define a value $a(i)$ as an average distance to elements within A:

$$a(i) = \frac{1}{|A|-1} \sum_{j \in A, j \neq i} d(i,j) \ .$$

For any cluster C different from A we define $d(i)$ as an average distance of i to objects in C:

$$d(i,C) = \frac{1}{|C|} \sum_{j \in C} d(i,j)$$

For any cluster C different from A we define $b(i)$ as the average distance to the members of the closest cluster:

$$b(i) = \min_{C \neq A} \{d(i,C)\}$$

Finally, the *silhouette value* $s(i)$ of the object i is defined as:

$$s(i) = \frac{b(i) - a(i)}{\max\{a(i), b(i)\}}$$

The silhouette value $s(i)$ for each object i lies between −1 and 1. If $s(i)=1$, the object is well classified; if $s(i)<0$, the object is badly classified (in fact, it is on average closer to members of some other cluster than the one to which it is assigned). The average silhouette value for a cluster can be used as a measure of the "goodness" of that cluster. The silhouette value characterizes not only the "tightness" of the given cluster, but also how far each element of the cluster is from the next closest cluster.

3.3 Supervised Analysis

One of the goals of supervised expression data analysis is to construct classifiers, such as linear discriminants, decision trees or support vector machines (SVM), which assign a predefined class to a given expression profile. For instance, if a classifier can be constructed based on gene expression profiles that are able to distinguish between two different but morphologically indistinct tumor tissues, such a classifier may be used in diagnostics. Moreover, if such a classifier is based on a set of relatively simple rules, it can help elucidate the mechanisms involved in the development of each tumor. Typically, such classifiers are trained on a subset of data with a-priori-given classification and tested for accuracy on another subset with known classification. After assessing the quality of the prediction they can be applied to new data, the classification of which is unknown.

Note that when classifying samples, we are confronted with a problem that there are many more attributes (genes) than objects (samples) that we are trying to classify (typically, there are around 10,000 genes and less than 100 samples). This makes it always possible to find a perfect discriminator if we are not careful in restricting the complexity of the permitted classifiers. To avoid this problem we must look for very simple classifiers that compromise between the simplicity and the classification accuracy. This is sometimes overlooked when elaborate classifiers are built for sample classification.

3.4 Data Visualization

An important aspect of all the analysis methods is the need to summarize the analysis results in a comprehensive way for human interpretation. Visualization techniques are used for looking at the large amounts of data simultaneously. One of the most popular techniques is the color-coding of gene expression matrices developed by Eisen et al. (1998). This "two-and-a-half"-dimensional visualization is quite powerful for even very large data sets and it can be used in conjunction with any clustering method.

Three-dimensional methods, as for example the visualization of the first three principal components of the expression data, can be used for rapid detection of outliers or large groups of similar objects. As the gene expression data are of very high dimensionality, consisting of measurements for tens of thousands of genes in hundreds or thousands of experiments, the mapping and visualization of a few principal components in 3-D have yet to prove its scalability for these very complex data.

Another 3-D method is the visualization of the "density landscape", a type of 2-D histogram where two dimensions represent the objects in a manner preserving spatial proximity in 2-D. VxInsight (Davidson et al. 1998) overlays the 2-D plane with a 3-D virtual landscape that looks like a mountain range. The height of a mountain is proportional to the density of objects beneath it. Users can navigate through this 3-D space and zoom in to identify individual objects and labels. This method allows rapid identification of the highest density areas, but it does not allow for easy representation of the actual expression measurements.

3.5 Tools

Many tools for gene expression matrix analysis have been developed. Tools developed at Stanford University – Cluster, XCluster and TreeView – are among the most popular gene expression matrix analysis tools that are free for academic use (see http://rana.stanford.edu/software). Array Explorer by Spotfire and GeneSpring by Silicon Genetics are among many of the available commercial tools for gene expression data analysis. We will not discuss any of these tools in detail here, since they are developing extremely fast, and new tools are coming onto the market. (An annotated index of the tools can be found at http://www.ncgr.org/research.genex/other_tools.html and www.microarray.org.)

4
Expression Profiler

Expression Profiler is a set of Internet tools aimed at the analysis and visualization of the gene expression data. The analysis of gene expression profiles can be combined with other types of data, and links to other databases and analysis tools over the Internet can be established by users themselves. While the tools described in the previous subsection have to be installed on the researcher's own computer, Expression Profiler offers a web-based service (see http://www.ebi.ac.uk/microarray), and the only requirement on the client's side is a web browser. Expression Profiler consists of four major components:

1. EPCLUST – "Expression Profile data CLUSTering and analysis", a collection of clustering and visualization methods for analysis of expression data.
2. URLMAP – a tool for mapping HTML form contents (e.g. cluster contents from EPCLUST) to other on-line analysis tools and databases.
3. GENOMES – a tool for simultaneous retrieval of information about sets of genes, links to other databases, and extraction of genomic sequences relative to the gene start and end positions.
4. SPEXS – "Sequence Pattern Exhaustive Search", a pattern discovery tool allowing for rapid exhaustive enumeration of all patterns occurring in sets of sequences, and reporting the most frequent, or over- or underrepresented patterns.

EPCLUST, the expression profile clustering and analysis tool, allows users to perform cluster analysis and visualization of expression data. The main methods available for analysis include efficient implementations of the hierarchical and *K*-means clustering approaches with various distance measures and clustering parameters (e.g. choice of distance measure, choice of hierarchical clustering method, choice of different ways to calculate the starting points for *K*-means etc.). Visualization of expression profiles is provided by the pseudo-coloring-based method developed by M. Eisen (Eisen et al. 1998). The convenience of the interactive usage and the speed of the algorithms should make it possible for investigators to explore the existing and new data becoming available from an increasing number of laboratories.

The clustering of gene expression data is only the first step of many in the potential analysis of gene expression data. GENOMES, the tool for retrieving the gene annotations and sequences, demonstrates the need for species-specific databases to be able to handle queries for sets of genes simultaneously. Users of expression data analysis would like to have "executive summaries" about the clusters they have discovered from the expression data. The new emerging genome-specific databases should also allow users to perform unconventional queries, for instance to extract the sets of upstream regulatory sequences for the set of genes in the query.

By using the organism-specific gene annotations and sequence download options, it becomes possible to integrate the sequence pattern discovery tools so that discovery of putative regulatory signals becomes part of the analysis pipeline. The only resource needed at the researcher's site is a computer with a web browser and an Internet connection.

URLMAP allows users to submit the analysis results, e.g. the cluster of co-expressed genes, to various other tools and databases for querying. One can, for example, ask the KEGG database, "In which metabolic pathways are the genes from my cluster participating?". Ideally, each on-line database should provide mechanisms to allow queries about many genes simultaneously, as opposed to the current quite common situation that restricts the querying to only one gene at a time.

The discovery of putative regulatory signals and transcription factor binding sites (Vilo et al. 2000) is an application of microarray expression data for gaining new biological knowledge. A pattern discovery tool, SPEXS, allows one to perform a rapid exhaustive search for a priori unknown, statistically significant sequence patterns of unrestricted length. The statistical significance is determined for a set of sequences in respect to a set of background sequences. This allows one to detect subtle regulatory signals specific for each cluster of co-expressed genes in comparison to the background distribution of all upstream sequences.

The four tools (EPCLUST, GENOMES, URLMAP and SPEXS) form currently the core of the Expression Profiler analysis suite. The tools can be used in the manner facilitating an automatic discovery of potential regulatory signals in genomes (Vilo et al. 2000), as briefly described in Section 6.

5 Applications

A simple method for finding sets of interesting genes is by comparing expression profiles of two or more samples in order to identify differentially expressed genes. For instance, Lee et al. (1999) used this method to find genes that are differentially expressed in skeletal muscle of adult (5-month) and old (30-month) mice. From over 6,347 mice genes surveyed using microarrays, 58 displayed a greater than two-fold increase, whereas 55 displayed a greater than two-fold decrease in expression in the skeletal muscles of the old mice. Of the genes that increased the expression, 16% were mediators or stress-response genes and 9% were involved in neuronal growth. Of genes that decreased in expression, 13% were participating in energy metabolism. In the same study, gene expression profiles from 30-month-old mice with restricted calorie intake (76% of that of control population) were compared with a 30-month-old control population, and it was shown that the expression profile of restricted-calorie-intake mice was closer to that of younger mice.

Clustering of expression profiles is now used routinely for grouping genes and samples with similar characteristics. The clustering of genes for finding co-regulated and functionally related groups is particularly interesting in the cases when we know the complete set of an organism's genes. In a seminal paper, DeRisi et al. (1997) used a DNA array containing the complete set of yeast genes to study the time course of the diauxic shift. They selected small groups of genes with similar expression profiles and showed that these genes are both functionally related and contain relevant transcription factor binding sites upstream of their open reading frames. More systematic studies of this data set for regulatory elements can be found in (Brazma et al. 1998; van Helden et al. 1998).

Later, more expression studies of yeast under various conditions were carried out, including sporulation (Chu et al. 1998), cell cycle (Spellman et al. 1998) and yeast gene regulatory machinery (Holstege et al. 1998). Clustering has been applied to the gene expression matrices, and groups of functionally related and co-regulated genes have been identified. Tavazoie et al. (1999) clustered 3,000 expression profiles of the most variable yeast genes during the cell cycle (15 time points; data from Cho et al. 1998) into 30 clusters using the *K*-means algorithm. They found that for half of these clusters, strong sequence patterns are present in the gene upstream sequences. The expression profiles of cell cycle-dependent genes should be periodic and Fourier analysis has been used to try to identify these genes (Spellman et al. 1998). Eisen et al. (1998) applied hierarchical clustering to gene expression matrices obtained by combining 80 different yeast samples (from different experimental conditions) studied in various hybridization experiments at Stanford University (including the ones mentioned above).

Similar approaches have been used for data from organisms other than yeast. For instance, Iyer et al. (1999) studied 8,600 genes in human fibroblast and obtained 10 distinct gene clusters each associated with genes with particular functional roles, such as signal transduction, coagulation and hemostasis, inflammation etc.

Hierarchical clustering has been used for sample clustering in a paper by Alizadeh et al. (2000), concretely for clustering of tumor samples by their expression patterns to find new possible tumor subclasses. Diffuse large B-cell lymphoma (DLBCL) has been studied using 96 samples of normal and malignant lymphocytes. Applying the hierarchical clustering (Eisen et al. 1998) to these samples, they showed that there is a diversity in gene expression among the tumours of DLBCL patients. They identified two molecularly distinct forms of DLBCL, which had gene expression patterns indicative of different stages of B-cell differentiation. Interestingly, these two groups correlated well with the patient survival rates, thus confirming that the clusters are meaningful.

The sample clustering approach has been combined with gene clustering to identify which genes are the most important for the sample clustering (Alon et al. 1999; Alizadeh et al. 2000). Alon et al. (1999) applied a partitioning-based clustering algorithm to study a gene expression matrix comprising 6,500 genes in 40 tumourous and 22 normal colon tissues and clustering by both genes and samples. They call this method two-way clustering.

Supervised data analysis has been used by Brown et al. (2000), who applied various supervised learning algorithms to six functional classes of yeast genes using gene expression matrices from 79 samples (Eisen et al. 1998). Genes from some of the classes, such as ribosomal proteins and histones, are expected to be co-expressed. For these classes a good classification accuracy is achieved. Some other functional classes, such as protein kinases, are not expected to have distinct gene expression profiles. It was shown that the SVM provides the best prediction accuracy for the functional classes that are expected to be co-regulated. When using supervised analysis methods, it would be desirable that the used method is not based on a "black box" concept where the discovered rules cannot be extracted or interpreted by humans.

Golub et al. (1999) applied neighbourhood analysis to construct class predictors for leukemia samples. They were looking for genes whose expression is best correlated with two known classes of leukemias – acute myeloid leukemia and acute lymphoblastic leukemia. They constructed a classifier based on 50 genes (from 6,817) using 38 samples and applied it to a collection of 34 new samples. The classifier correctly predicted 29 of these 34 samples. Note that in fact two genes are enough to build a linear discriminant perfectly classifying these tumor samples (M. Niranjan, pers. comm. 2000).

Ben-Dor et al. (2000) applied a new clustering algorithm for classification of colon and ovarian cancer data sets. They used unsupervised clustering to find a hierarchical structure in the expression profile space, and supervised learning to find the best threshold to correlate the clustering structure with the known cancer classes.

6
From Gene Expression Data to Gene Regulatory Networks

Whether we use supervised or unsupervised expression profile analysis, or extensive data visualizations, these are only the first steps in expression data analysis. It is a long way from finding gene clusters to finding the functional roles of the respective genes, and moreover, to understanding the underlying biological processes. A natural step after clustering the expression profiles is to study the putative promoter sequences of similarly expressed genes to find potential regulatory sequence elements in genomes. This is relatively easy for yeast, since typically yeast promoters are relatively close to the beginning of open reading frames (ORFs). In this section we describe an approach that uses gene expression data to find regulatory sequence elements in yeast.

A major role in gene regulation in eukaryotic organisms is played by specific proteins, called *transcription factors*. By binding to sequence-specific sites in the DNA, called *transcription factor binding sites*, they influence the transcription of a particular gene. The transcription factor binding sites are located in *promoter regions*. In yeast these regions are predominantly (but not exclusively) in the immediate vicinity of the gene (typically less than 1,000 bp upstream of the translation start site).

It seems reasonable to hypothesize that genes with similar expression profiles, i.e. genes that are *co-expressed*, may have common regulatory mechanisms, i.e. they may be *co-regulated*, and hence have similar transcription factor binding sites. Therefore by clustering together genes with similar expression profiles one can find groups of potentially co-regulated genes, allowing one to search for putative regulatory signals in their upstream regions.

During the last couple of years, the outlined approach has been widely used. For instance, DeRisi et al. (1997) studied the diauxic shift in yeast. They found several distinct clusters in the gene expression profiles and were able to show the presence of several transcription factor binding sites located upstream of the respective genes. The data set by DeRisi et al. (1997) was also studied by Van Helden et al. (1998), who systematically searched for oligonucleotides over-represented upstream of potentially co-regulated genes and showed that potential transcription factor binding sites can be found in this way. A systematic analysis of over-represented sequence patterns in clusters of upstream sequences obtained by clustering the diauxic shift expression profiles has been carried out by Brazma et al. (1998). It was shown that over-represented patterns present in sequences upstream of genes from expression profile clusters of at least 25 genes cannot be explained by statistical "chance", i.e. they are statistically significant and may have biological relevance. Many of the discovered patterns have matches in known yeast transcription binding site descriptions.

More expression studies have been carried out recently under various conditions (e.g. sporulation and cell cycle; Chu et al. 1998; Spellman et al. 1998) and the number of such studies is increasing rapidly. Tavazoie et al. (1999) clustered 3,000 expression profiles of the most variable yeast genes during the cell cycle (15 time points; data from Cho et al. 1998) into 30 clusters by the *K*-means algorithm. They found that for half of these clusters, strong sequence patterns are present in upstream sequences. Moreover, it was noted that upstream sequences from genes from the tightest expression profile clusters (i.e. the clusters with smaller average Euclidean distance to their centers) contain more significant patterns. In a different study, Wolfsberg et al. (1999) studied the discovery of potential regulatory sequences from groups of genes whose expression was peaked at distinctive phases of cell cycle as identified by Cho et al. (1998). Cell cycle data have also been studied by Zhang (1999) and van Helden et al. (2000). Jensen and Knudsen (2000) used suffix tree-based substring discovery methods for identifying putative binding sites in *Saccharomyces cerevisiae*. Their method, however, was not based on the clustering of expression profiles. Instead, they sorted genes based on the expression measurements from individual time points and used Kolmogorov-Smirnov tests for identifying the co-regulated genes and putative binding sites.

Most of the authors concentrate in their studies on the analysis of only a few carefully chosen gene expression clusters and possibly on a restricted pattern representation language. A most systematic in silico regulatory element discovery experiment would be the following:

1. Cluster the genes based on a selection of expression measurements.
2. Extract putative promoter sequences for the genes in each cluster.
3. Search for sequence patterns over-represented in these clusters.
4. Assess the quality of discovered patterns using a statistical significance criteria.

Such an approach has been implemented in the study by Vilo et al. (2000). Over 52,000 different clusterings of gene expression profiles for yeast (80 experimental conditions taken from Stanford data set) have been created by *K*-means clustering and the "goodness" of each cluster has been assessed using an average silhouette measure (see Sect. 3.2). For each of the clusters an exhaustive search of statistically over-represented sequence patterns has been done in genome sequences upstream of the respective genes. About 1,500 significant patterns have been selected by formal criteria and matched against the experimentally mapped transcription factor binding sites in SCPD database. The 1500 patterns have been clustered in 62 groups for which alignments and consensus patterns have been derived. Of these 62 groups, 48 had patterns that have matching sites in the SCPD database. The others are potentially novel regulatory signals in the yeast genome.

To reduce the number of patterns, the hierarchical clustering method (the same as used for clustering of expression profiles) based on the pairwise distances between all patterns has been used. The tree produced by hierarchical clustering can be cut using different thresholds in order to get clusters of similar patterns. For each such cluster, alignments as well as consensus patterns are built. This method reduces the number of patterns from 1,500 to about 50–200 depending on the threshold. This number of patterns seems to be reasonable for further human inspection and potential wet-lab verifications.

Examples of the discovered highly rated patterns that were not present in SCPD or TRANSFAC databases are sites GGTGGCAA and GGTGGCAAA. These patterns were discovered from clusters of genes, where most of the genes were proteasome subunits. The same pattern, a binding site for protein Rpn4p, has been independently reported by other researchers (Jensen and Knudsen 2000; M. Eisen, pers. comm. 2000), as well as verified by experimental evidence (Mannhaupt et al. 1999).

7 Conclusions

Currently, the experiments on gene expression profiling are analogous to the way gene sequencing was done before the era of genome sequencing: they are mostly carried out to study a particular problem or sometimes just to demonstrate the concept. Over time, with the microarray technology becoming more reliable, introduction of standard controls and data normalization methods, it will become possible to systematically profile genes in various organisms, cell types, developmental stages and conditions. Various chemical compounds will be profiled for their effects, such as toxicity, on organisms and the results stored in databases. This approach will resemble systematic genome sequencing.

However, there is a major difference between genome sequence and expression data. Even if eventually we are able to overcome various technological limitations, and even if we are able to measure gene expression in terms of absolute units, such as mRNA counts, the gene expression profiles are meaningful only in the context of the experimental conditions in which they have been measured. This requires detailed and systematic annotation of samples and experimental conditions. For this to become a reality, agreed ontologies and controlled vocabularies for tissues, cell types and treatments, as well as for array designs, image analyses and hybridization protocols, have to be developed. Efforts to co-ordinate such standardization are underway (see www.mged.org).

Another necessary prerequisite for systematic gene expression profiling is the establishment of public repositories for gene expression data (Brazma et al. 2000). Integration of these databases with other genomic databases, as well as with analysis and visualization tools, will become a necessity. Different types of data need to be integrated with gene expression data in order to understand the mechanisms of living cells. For example, the sequence data (protein coding sequences, upstream regulatory sequences), protein-protein interactions, metabolic pathways and other types of information provide us with valuable information.

New challenges will arrive, as researchers become more interested in information about potentially large sets of genes, rather than in individual genes. This means that databases should provide overview answers about the similarities and dissimilarities of genes in an expression cluster. These "executive summaries" will be important to gaining a better understanding of how groups of genes behave collectively. Analysis suites for the analysis of vast amounts of microarray gene expression data will become an essential part of microarray expression databases.

Expression data analysis methods are currently only in their infancy. Even the rather obvious approaches, such as cluster analysis and finding differentially expressed genes, have been used only rather crudely. For instance, the appropriateness of similarity measures has not been explored systematically and these measures are typically used in an ad-hoc manner. The information characterizing the quality or reliability of different data points is not used typically. In the next generation of microarrays, where each spot is printed or synthesized many times, it will be much easier to estimate the measurement reliability by the standard deviation between the individual measurements.

Similarly to genome sequencing, a systematic gene expression profiling is not an end in itself, but is only creating an infrastructure for further research. It is a long way from having detailed gene expression profiles to real understanding of underlying cellular processes. Bioinformatics methods and tools will be needed to cope with the huge amounts of data, but they will not bring any deep understanding by themselves. On the other hand, the traditional "gene by gene" methods will not be sufficient to understand gene regulatory networks consisting of thousands or tens of thousands of genes. Hypothesis-driven and data mining approaches will be used hand in hand with high-throughput data analysis.

References

Alizadeh AA, Eisen MB, Davis RE, Ma C, Lossos IS, Rosenwald A, Boldrick JC, Sabet H, Tran T, Yu X, Powell JI, Yang L, Marti GE, Moore T, Hudson J Jr, Lu L, Lewis DB, Tibshirani R, Sherlock G, Chan WC, Greiner TC, Weisenburger DD, Armitage JO, Warnke R, Levy R, Wilson W, Grever MR, Byrd JC, Botstein D, Brown PO, Staudt LM (2000) Distinct types of diffuse large B-cell lymphoma identified by gene expression profiling. Nature 403:503–511

Alon U, Barkai N, Notterman DA, Gish K, Ybarra S, Mack D, Levine AJ (1999) Broad patterns of gene expression revealed by clustering analysis of tumor and normal colon tissues probed by oligonucleotide arrays. Proc Natl Acad Sci USA 96:6745–6750

Ben-Dor A, Bruhn L, Friedman N, Nachman I, Schummer M, Yakhini Z (2000) Tissue classification with gene expression profiles. In: Shamir R, Miyano S, Istrail S, Pevzner P, Waterman M (eds) Proc 4th Annu Int Conf on Computational Molecular Biology RECOMB-2000, Tokyo, Japan. ACM Press, New York, pp 54–64

Ben-Dor A, Shamir R, Yakhini Z (1998) Clustering gene expression profiles. J Comput Biol 6(3–4):281–297

Brazma A, Jonassen I, Vilo J, Ukkonen E (1998) Predicting gene regulation elements in silico on a genomic scale. Genome Res 8:1202–1215

Brazma A, Robinson A, Cameron G, Ashburner M (2000) One stop shop for microarray data. Nature 403:699–700

Brown MPS, Grundy WN, Lin D, Cristianini N, Sugnet CW, Furey TS, Ares M Jr, Haussler D (2000) Knowledge-based analysis of microarray gene expression data by using support vector machines. Proc Natl Acad Sci USA 97:262–267

Cho RJ, Campbell MJ, Winzeler EA, Steinmetz L, Conway A, Wodicka L, Wolfsberg TG, Gabrielian AE, Landsman D, Lockhart DJ, Davis RW (1998) A genome wide transcriptional analysis of gene expression of the mitotic cell cycle. Mol Cell 2:65–73

Chu S, DeRisi JL, Eisen M, Mulholland J, Botstein D, Brown PO, Herskowitz I (1998) The transcription program of sporulation in budding yeast. Science 282:699–705

Claverie J-M (1999) Computational methods for the identification of differential and coordinated gene expression. Hum Mol Genet 8(10):1821–1832

Davidson GS, Hendrickson B, Johnson DK, Meyers CE, Wylie BN (1998) Knowledge mining with VxInsight: discovery through interaction. J Intelligent Info Syst Integrating Artificial Intelligence Database Technol 11(3):259–285

DeRisi JL, Iyer VR, Brown PO (1997) Exploring the metabolic and genetic control of gene expression on a genomic scale. Science 278:680–686

D'haesleer P, Wen X, Fuhrman S, Somogyi R (1998) Mining the gene expression matrix: Inferring gene relationships from large scale gene expression data. In: Paton RC, Holcombe M (eds) Information processing in cells and tissues. Plenum, London, pp 203–212

Duggan D, Bittner M, Chen Y, Meltzer P, Trent J (1999) Expression profiling using cDNA microarrays. Nat Genet 21 (Suppl):10–15

Eisen M, Spellman PT, Botstein D, Brown PO (1998) Cluster analysis and display of genome-wide expression patterns. Proc Natl Acad Sci USA 95:14863–14867

Golub TR, Slonim DK, Tamayo P, Huard C, Gaasenbeek M, Mesirov JP, Coller H, Loh ML, Downing JR, Caligiuri MA, Bloomfield CD, Lander ES (1999) Molecular classification of cancer: class discovery and class prediction by gene expression monitoring. Science 286:531–537

Hartigan JA (1975) Clustering algorithms. Wiley, New York

Holstege FC, Jennings EG, Wyrick JJ, Lee TI, Hengartner CJ, Green MR, Golub TR, Lander ES, Young RA (1998) Dissecting the regulatory circuitry of a eukaryotic genome. Cell 95(5):717–728

Iyer VR, Eisen MB, Ross DT, Schuler G, Moore T, Lee JCF, Trent JM, Staudt LM, Hudson J Jr, Boguski MS, Lashkari D, Shalon D, Botstein D, Brown PO (1999) The transcriptional program in the response of human fibroblasts to serum. Science 283:83–87

Jain AK, Murty MN, Flynn PJ (1999) Data clustering: a review. ACM Comput Surv 31:264–323

Jensen LJ, Knudsen S (2000) Automatic discovery of regulatory patterns in promoter regions based on whole cell expression data and functional annotation. Bioinformatics 16(4):326–333

Lee CK, Klopp RG, Weindruch R, Prolla TA (1999) Gene expression profile of aging and its retardation by caloric restriction. Science 285:1390–1393

Lee M-LT, Kuo FC, Whitmore GA, Sklar J (2000) Importance of replication in microarray gene expression studies: statistical methods and evidence from repetitive cDNA hybridizations. Proc Am Assoc Sci 97:9834–9839

Legendre P, Legendre L (1998) Numerical ecology. Developments in environmental modelling. Elsevier, Amsterdam

Li M, Vitanyi P (1993) An introduction to Kolmogorov complexity and its applications. Springer, Berlin Heidelberg New York

Mannhaupt G, Schnall R, Karpov V, Vetter I, Feldmann H (1999) Rpn4p acts as a transcription factor by binding to PACE, a nonamer box found upstream of 26S proteasomal ad other genes in yeast. FEBS Lett 450:27–34

Rousseeuw PJ (1987) Silhouettes: a graphical aid to the interpretations and validation of cluster analysis. J Comput Appl Math 20:53–65

Sander C (2000) Genomic medicine and the future of health care. Science 287:197–198

Schuchhardt J, Beule D, Malik A, Wolski E, Eickhoff H, Lehrach H, Herzel H (2000 Normalization strategies for cDNA microarrays. Nucleic Acids Res 28(10):E47

Sharan R, Shamir R (2000) CLICK: a clustering algorithm with applications to gene expression data. In: Proc 8th Int Conf on Intelligent Systems for Molecular Biology AAAI Press, Menlo Park, California, pp 307–316

Spellman PT, Sherlock G, Zhang M, Iyer VR, Anders K, Eisen M, Brown PO, Botstein D, Futcher B (1998) Comprehensive identification of cell cycle-regulated genes of the yeast *Saccharomyces cerevisiae* by microarray hybridization. Mol Biol Cell 9:3273

Tamayo P, Slonim D, Mesirov J, Zhu Q, Kitareewan S, Dmitrovsky E, Lander ES, Golub TR (1999) Interpreting patterns of gene expression with self-organizing maps: methods and application to hematopoietic differentiation. Proc Natl Acad Sci USA 96(6):2907–2912

Tavazoie S, Hughes D, Campbell MJ, Cho RJ, Church GM (1999) Systematic determination of genetic network architecture. Nat Genet 22:281–285

Tibshirani R, Hastie T, Eisen M, Ross D, Botstein D, Brown P (1999) Clustering methods for the analysis of DNA microarray data. Technical Rep, Department of Statistics, Stanford University, Stanford

Toronen P, Kolehmainen M, Wong G, Castren E (1999) Analysis of gene expression data using self-organizing maps. FEBS Lett 451(2):142–146

van Helden J, Andre B, Collado-Vides J (1998) Extracting regulatory sites from the upstream region of yeast genes by computational analysis of oligonucleotide frequencies. J Mol Biol 281(5):827–842

van Helden J, Andre B, Collado-Vides J (1998) Extracting regulatory sites from the upstream region of yeast genes by computational analysis of oligonucleotide frequencies. J Mol Biol 281(5):827–842 van Helden J, Rios AF, Collado-Vides J (2000) Discovering regulatory elements in non-coding sequences by analysis of spaced dyads. Nucleic Acids Res 28(8):1808–1818

Vilo J, Brazma A, Jonassen I, Robinson A, Ukkonen E (2000) Mining for putative regulatory elements in the yeast genome using gene expression data. In: Proc 8th Int Conf on Intelligent Systems for Molecular Biology. AAAI Press, Menlo Park, California, pp 384–394

Wolfsberg TG, Gabrielian AE, Campbell MJ, Cho RJ, Spouge JL, Landsman D (1999) Candidate regulatory sequence elements for cell cycle-dependent transcription in *Saccharomyces cerevisiae.* Genome Res 9(8):775–792

Young R (2000) Biomedical discovery with DNA arrays. Cell 102:9–16

Zhang MQ (1999) Promoter analysis of coregulated genes in the yeast genome. Comput Chem 23:233–250

CHAPTER 7

Future Trends in the Use of DNA Arrays for Expression Measurement

BERTRAND R. JORDAN[1]

[1] Marseille-Génopole, Parc Scientifique de Luminy, Case 901, 13288 Marseille, Cedex 9, France

1 Introduction

Futurology is always a very dangerous exercise: predictions are almost certain to be wrong in some respect. On the other hand, it is indeed useful to try to discern likely trends, if only to orient decisions on the purchase of equipment. I will therefore discuss briefly, in this closing chapter, the directions in which I feel the DNA array field is likely to go during the next 2 or 3 years.

2 Future Chips: How Complex?

With the completion of the human (and other) sequences, the wish to assess the expression of all human genes in a single step becomes overwhelming, while the existence of sequence data and of the corresponding cloned segments simplifies the task. Thus the race to higher density towards a "whole-genome" chip or microarray allowing simultaneous measurement of the expression of all human genes will continue. Even if the actual number of human genes turns out to be 30,000 or 40,000 instead of the often quoted figure of 100,000 (Pennisi 2000), this will not be trivial. At this time, the highest reported densities for microarrays used in actual experiments are of the order of 3,000 spots/cm^2. An array covering most of the surface of a microscope slide, at this density, would display "only" 20,000 genes (without duplicates or controls). Thus the human or murine complement of genes is likely to be covered by a set of several microarrays rather than a single one. Increased density would not tax existing microarray scanners (10- or even 5-µm resolution is already current); the limitation lies mainly with spotting devices. It is quite difficult to consistently make spots of DNA solutions having a diameter of less than 100 µm using solid pins, split pins or piezoelectric devices.* Allowing for some spot separation, this translates into a "pitch" of around 200 µm, i.e. 2,500 spots/cm^2. Changes in spotting mechanisms and surface chemistry may allow closer spacing, but in any case the future of very complex microarrays is probably limited because of the difficulty and expense involved in producing such large numbers of PCR products (see below).

Concerning oligonucleotide chips, devices currently marketed by the firm Affymetrix contain nearly 400,000 short (20-mer) oligonucleotides on a single chip. However, because 30–40 oligonucleotides (including mismatched con-

---

* This corresponds approximately to 1 nl of solution. Current inkjet printers do produce much smaller drops (a few picolitres), but this is using specially formulated, extremely fluid and clean inks. DNA solutions do not have the required properties and in addition tend to vary from sample to sample.

trols) are used for each gene, the latest "Human Genome U95" set from this firm requires five arrays to assay approximately 62,000 genes. The relatively poor yield of the photochemical on-chip synthesis process used does not allow (so far) the manufacture of long oligonucleotides that would provide more specificity with fewer "features". A limited decrease in the number of oligonucleotides can be achieved through progress in methods allowing the choice of the "best" sequences for this purpose, but drastic reduction would jeopardise the quality of the measurement. On the other hand, increasing the chip density and placing one or a few million "features" on the surface of a microscope slide is probably feasible, although the resolution of reading devices would then have to be improved. In summary, assessing all human genes with one or two Affymetrix chips is likely to be eventually possible but to represent the limit of this technology. New on-chip synthesis approaches allowing the production of long oligonucleotides (see below) may remove this limitation.

In any case such full human genome chips will definitely not be cheap, and problems in data acquisition, storage and analysis are likely to be formidable, making smaller, specialised arrays attractive for many purposes.

3
From "Clone-Based" to "Sequence-Based Arrays"

Newcomers to the microarray field, after having assembled the necessary robotic equipment, often discover with dismay the difficulty, expense and tedium involved in assembling collections of thousands of cDNA clones and producing sufficient amounts of purified DNA from each of them by PCR. Many hurdles (errors in the IMAGE collection, contamination by bacteriophages) complicate the task. There are still only a few commercially available microarrays, presumably because of the same difficulties, compounded by intellectual property issues (see later). In addition, they are likely to remain quite expensive (GEM microarrays marketed by Incyte and allowing the assay of 10,000 genes are priced at US$4,000 apiece), even with increased production volumes: there are no great economies of scale to be expected in the manufacture of thousands of microarrays. Both the cost of producing the DNA segments by PCR and the actual expense involved in spotting the arrays increase almost linearly with the number of devices manufactured.

Oligonucleotide chips do not suffer from the same problem. They eliminate completely the recourse to clones since they are based solely on sequence information – which is already vast for many organisms and is increasing at an explosive rate. In addition, economies of scale can be considerable. Affymetrix, at present the dominant provider in this field, performs 80 (20×4) cycles of base addition reactions to build sets of 20-mer oligonucleotides on fairly large glass "wafers" from which the individual chips are then cut out. It cannot cost very much more to produce a set of 400 small, very high-density chips than to make the set of 49 larger, lower-density devices previously manufactured on the same glass wafer. Yet the new chips can carry as many, or even more, different oligonucleotides than the previous ones. These economies of scale are not yet strongly reflected in the company's price list (although they do appear in its willingness to negotiate large discounts for volume purchases), but this will change if competition increases.

A number of laboratories and firms are developing "on-chip" oligonucleotide synthesis techniques that rely, for example, on fast dispensing of synthesis reagents to individual sites on the chip by printhead-like devices; some of them have already begun to market their processes or products (Protogene, http://www.protogene.com; Oxford Gene Technologies, http://www.ogt.co.uk; Agilent, http://www.chem.agilent.com). Such procedures allow the use of classical synthesis chemistry (rather than the less efficient photochemical method), making possible the manufacture of much longer oligonucleotides (50- to 100-mers) which, in turn, reduce the need for redundancy in the chip because of their higher specificity. This makes it easier to represent many genes on a chip since only one or a few (long) oligonucleotides are needed to assay each of them. In addition, these approaches are inherently more flexible that the Affymetrix photochemical method as the manufacture of a different chip simply involves reprogramming the dispensing of reagents, rather than the manufacture of a complete new series of 80 (20×4) masks. Other firms (for example, DNAmicroarray, http://DNAmicroarray.com; Operon Technologies, http://geneop.operon.com; Mergen, http://www.mergen-ltd.com/prod.html; also Agilent, http://www.chem.agilent.com) offer arrays made with pre-synthesised (long) oligonucleotides.

The development of these technologies will not depend solely on scientific and engineering advances: intellectual property in this field is already a hotly contested issue. A series of patents issued to Affymetrix contain wide-ranging claims that, taken at face value, would prevent the manufacture and sale of almost any kind of high-density DNA array. Litigation is in progress, notably with Incyte; other manufacturers await the outcome.* These patents are contested, however, notably by E. Southern who first described the use of "multitudes of oligonucleotides tethered to a glass slide" in a 1989 paper at the Cold Spring Harbor Genome Mapping and Sequencing Meeting (Southern and Maskos 1989). Lawyers are having a field day with this contest; hopefully, it will be resolved in a fashion that opens up competition.

* The firm DNAmicroarray mentions on its web site "Although densities greater than 10,000 spots per centimeter squared are readily available, only up to 400 spots per centimeter squared is offered at this time. We as well as others are anticipating the court decision on the pending lawsuit filed by Affymetrix vs. Incyte".

Altogether, it is almost certain that ready-made, sequence-based DNA chips will be increasingly used in the future, certainly for the "standard" sets for which the prime example is the yeast whole-genome chip and, possibly, depending on methods development, for more specialised arrays.

4 From "Home-Made" Microarrays to Ready-Made Chips

Oligonucleotide chips discussed in the preceding section will not be made by research laboratories, but instead bought from companies. Even for clone-based microarrays, a shift towards purchase of manufactured devices is likely. It does not make economic sense for individual groups or even research institutes to invest large resources in the construction of standard microarrays (again, the yeast whole-genome microarray is a good example). This task can be handled more efficiently by industry or, in some cases, by public resource centres. This is not to say that microarray manufacture will disappear from the research environment: custom arrays allowing the assay of limited, specialised sets of genes will remain useful in many cases, and maximum flexibility can be achieved by making them "in house". Alternately, some manufacturers may undertake to produce such custom arrays, while others will provide sets of "ready-to-spot" PCR products: this is already offered by Incyte with its "Easy to spot" PCR products (http://www.incyte.com/reagents) and by Research Genetics with MyArray DNA (http://www.resgen.com/products). The end result is likely to be a mixed situation in which large or standard sets of genes are assessed with commercial oligonucleotide chips or microarrays, while custom arrays are made in various academic-corporate arrangements. Of course in this context it is very important to standardise detection systems so that each type of industry-produced DNA array does not require its own "proprietary" scanning device priced at US$ 50,000 to 100,000.

5 From "Stand Alone" Array to Integrated "Lab-on-a-Chip"

Biochip technology is not limited to DNA arrays. The integration of a number of functionalities within chips whose dimensions are measured in centimetres is well underway; such devices can perform filtration, fluid handling and reagent mixing, PCR reactions and even capillary electrophoresis (Talary et al. 1998; see http://www.chem.agilent.com/cag/products for one of the first marketed devices). Their development is strongly stimulated by the need of pharmaceutical companies to perform literally millions of tests in the course of screening compounds for activities ("high throughput screening"), and by the requirement to do these assays very quickly, in a highly parallel mode and with the smallest possible amount of reagents. At least for industrial and clinical systems, expression measurement (probably assessing limited numbers of genes) is likely to be packaged in with such systems. This is, for example, the form in which expression measurement will penetrate in clinical oncology laboratories – if indeed the clinical utility of such data is confirmed (Alizadeh et al. 2000; Bertucci et al. 2000).

6
From Fluorescent Labelling to Electrical Detection

Fluorescent labelling is relatively cumbersome, interferes by steric hindrance with hybridisation, and requires high-end, expensive detection systems; radioactive labelling is undesirable in many environments, and provides limited resolution even with high-performance (and costly) detectors. It is very desirable to achieve detection of the fact that a given location in the array has hybridised, and to quantify the extent of hybridisation, by some other method. This should preferably involve the measurement of an electrical signal, and would, ideally, require no modification of the sample before hybridisation. Much effort is devoted by many groups towards achieving this (Souteyrand et al. 1997; Wang et al. 1999). The approaches explored range from the detection of some subtle change of electrical properties upon hybridisation, to very exotic methods: microbalances "weighing" the extra mass of the hybridised material, or determination of the number of double-stranded (thus hybridised) molecules by atomic force microscopy. Proof of principle has been obtained for some of these approaches; it remains to be seen whether they can achieve the required sensitivity and throughput. If successful, they are likely to have an impact first in applications of DNA arrays such as bacterial identification or mutation detection, where a "yes/no" answer is often sufficient, rather than in expression measurement where accurate quantification is required.

7
Progress Towards More Sophisticated Data Interpretation and (Maybe) a Unified Expression Repository

Software and bioinformatics development is a very important aspect that was not sufficiently taken into account at the beginning of the "DNA array revolution". Even today, the type of analysis performed on expression data remains relatively unsophisticated; in addition, many of the actual data are still unavailable outside of the originator's laboratory and the selected data sets provided by some groups on their web sites lack a common format, making them directly usable by others. As discussed elsewhere in this book (Chap. 6), great efforts are being made to develop better analysis software, including both extensive statistical, correlation and clustering analysis, and direct links to current, constantly updated information available on the web. In addition, serious attempts are made to define a standard data format that would make it possible to store expression data in the way in which DNA sequences have been archived, and to make it thus generally available and useful to the research community. A number of data repositories already exist (for up-to-date lists see http://www.ncgr.org/research/genex/other_tools.html and http://www.biologie.ens.fr/en/genetiqu/puces/bddeng.html), but so far there is no unified system comparable to the GenBank and EMBL sequence databases. Of course, the problem of data format and standardisation is much more complex for expression data than for sequence information.

8 Expression Measurement Is Here to Stay

This is an easy prediction to make. Undoubtedly, other methods likely to add functional significance to gigabases of DNA sequence will be streamlined, made more efficient and more amenable to large-scale implementation: protein interaction studies, proteomics in general, gene inactivation studies in various model systems are bound to become faster, easier, cheaper. However, large-scale expression measurement, enhanced by general availability of sequence data and boosted by technical development of DNA arrays, will certainly remain a major approach in biology for quite a long time.

References

Alizadeh AA, Eisen MB, Davis RE, Ma C, Lossos IS, Rosenwald A, Boldrick JC, Sabet H, Tran T, Yu X, Powell JI, Yang L, Marti GE, Moore T, Hudson J Jr, Lu L, Lewis DB, Tibshirani R, Sherlock G, Chan WC, Greiner TC, Weisenburger DD, Armitage JO, Warnke R, Staudt LM, et al. (2000) Distinct types of diffuse large B-cell lymphoma identified by gene expression profiling. Nature 403:503–511

Bertucci F, Houlgatte R, Benziane A, Granjeaud S, Adelaide J, Tagett R, Loriod B, Jacquemier J, Viens P, Jordan B, Birnbaum D, Nguyen C (2000) Gene expressoin profiling of primary breast carcinomas using arrays of candidate genes Hum Mol Gene (in press)

Pennisi E (2000) Human Genome Project. And the gene number is...? Science 288:1146–1147

Souteyrand E, Cloarec JP, Martin JR, Wilson C, Lawrence I, Mikkelsen S, Lawrence MF (1997) Direct detection of the hybridization of synthetic homo-oligomer DNA sequences by field effect. J Phys Chem B101:2980–2985

Southern EM, Maskos U (1989) Synthesis of oligonucleotides tethered to a glass surface: applications in the analysis of nucleic acid sequences. Genome Mapping and Sequencing Meeting, Cold Spring Harbor, p 136

Talary MS, Burt JP, Pethig R (1998) Future trends in diagnosis using laboratory-on-a-chip technologies. Parasitology 117 (Suppl):S191–S203

Wang J, Jiang A, Mukherjee B (1999) New label-free DNA recognition based on doped nucleic-acid probes within conducting polymer films. Anal Chim Acta 402:7–12

Subject Index